Materials Science and Technologies

nova science publishers

www.novapublishers.com

Materials Science and Technologies

High-Performance Calcium-Carbonate Concrete
Natt Makul (Author)
2022. ISBN: 978-1-68507-412-8 (eBook)

Analytical Models of Interstitial-Atom-Induced Stresses in Isotropic Metallic Materials
Ladislav Ceniga (Editor)
2022. ISBN: 978-1-68507-429-6 (eBook)

Analytical Models of Coherent-Interface-Induced Stresses in Composite Materials II
Ladislav Ceniga (Author)
2021. ISBN: 978-1-68507-003-8 (eBook)

Analytical Models of Coherent-Interface-Induced Stresses in Composite Materials III
Ladislav Ceniga (Author)
2021. ISBN: 978-1-53619-996-3 (eBook)

Microgels: Synthesis, Properties and Applications
Wing-Fu Lai (Editor)
2018. ISBN: 978-1-53613-522-0 (Hardcover)
2018. ISBN: 978-1-53613-523-7 (eBook)

Zirconia and Alumina Bioceramic Biocompatibility
Giulio Maccauro, Pierfrancesco Rossi Iommetti, Luca Raffaelli and Paolo Francesco Manicone (Editors)
2012. ISBN: 978-1-62081-650-9 (Online Book)

More information about this series can be found at https://novapublishers.com/product-category/series/materials-science-and-technologies/

Alina Iuliana Pruna
Editor

Aerogels

Properties and Applications

nova
science publishers
www.novapublishers.com

DOI: https://doi.org/10.52305/WJOG8919

NOTICE TO THE READER

Additional color graphics may be available in the e-book version of this book.

Library of Congress Cataloging-in-Publication Data

Names: Pruna, Alina Iuliana, editor.
Title: Aerogels : properties and applications / [edited by] Alina Iuliana Pruna, professor, materials science, Universitat Politecnica de Valencia, Valencia, Spain.
Other titles: Aerogels (Nova Science Publishers)
Description: New York : Nova Science Publishers, [2022] | Series: Materials science and technologies | Includes bibliographical references and index. |
Identifiers: LCCN 2022018836 (print) | LCCN 2022018837 (ebook) | ISBN 9781685077884 (hardcover) | ISBN 9781685078959 (adobe pdf)
Subjects: LCSH: Aerogels. | Aerogels--Industrial applications.
Classification: LCC TP862 .A36 2022 (print) | LCC TP862 (ebook) | DDC 541/.34514--dc23/eng/20220720
LC record available at https://lccn.loc.gov/2022018836
LC ebook record available at https://lccn.loc.gov/2022018837

Published by Nova Science Publishers, Inc. † New York

Dedication

I dedicate this book to my family for their unconditioned love and support. I also thank my friend, N, for the suggestions and support to make this project successful.

Contents

Preface

Aerogels are three-dimensional architectures of interconnected porous networks exhibiting extra-low density and high active surface area that are usually obtained from their hydrogel form by drying techniques. The extraordinary properties of varying aerogels including their porous structure, specific surface area have attracted increasing interest for improving device performance.

The book entitled 'Aerogels: Synthesis and Applications' aims to provide with an insight into the synthesis and applications of varying materials in their aerogel form. The book covers the recent advances and reviews the state of the art of aerogels prepared from varying nanomaterials including carbon nanomaterials, silica, bioaerogels and hybrid aerogels in applications such as energy storage, electronics, or environmental remediation -in term of water and air.

The book is thought to serve as a fundamental information source on the techniques and methodologies applied to obtain the aerogel form of different nanomaterials, their characterization, their post-synthesis modification or hybridization. The information enclosed in this book will be invaluable to the experts to consolidate their knowledge and provide with an insight also for the newcomers to the field. Moreover, the book also includes the evolution of student satisfaction regarding a novel teaching approach based on the use of a teaching mobility event (International Week) as a platform to introduce engineering students to Nanotechnology and specifically to graphene aerogels which could serve in the field for improving teaching abilities and outcomes.

The book consists in 8 chapters as detailed below:

1) In the first chapter, Almeida-Naranjo and Guerrero introduce the readers to different aerogels employed as adsorbents for water treatment.

2) In the second chapter, Rosas-Laverde and coworker provide with the recent research in water remediation from oil spills by using graphene aerogels obtained by templated and non-templated methods.
3) In the third chapter, Khanuja and coworkers present the application of aerogels for photocatalytic degradation of air and water pollutants such as dyes, heavy metal ions, etc.
4) In the fourth chapter, Capulin presents the aerogels applied for thermal and acoustic insulation.
5) In the fifth chapter, Garcia Ruiz presents a state of the art on the synthesis methods of silica aerogels and their further surface modification techniques.
6) In the sixth chapter, Galeas Hurtado presents the characterization techniques applied for aerogels and the performance of aerogel-based devices in biosensing field including glucose sensing, antigen sensing, contaminants etc.
7) In the seventh chapter, Pruna and coworkers reviews the recent advances in the aerogels based on carbon nanomaterials namely carbon nanotubes, graphene and nanodiamonds and their applications.
8) In the eighth and last chapter, Pruna presents the student satisfaction degree on a novel teaching approach based on the exploitation of a teaching mobility event (International Week) to introduce engineering students to Nanotechnology and specifically to graphene aerogels.

Acknowledgments

I would like to warmly thank my colleagues for their interesting inputs for this book project.

Chapter 1

Aerogels for Contaminant Removal from Water

Cristina E. Almeida-Naranjo[1,*] and Victor H. Guerrero[2]

[1]Department of Mechanical Engineering, Escuela Politécnica Nacional, Quito, Ecuador
[2]Department of Materials, Escuela Politécnica Nacional, Quito, Ecuador

Abstract

The continuous deterioration of water quality and the increasing pressures on water resources can have serious effects on social development, human health and the environment. Therefore, it is necessary to formulate strategies and develop technologies for responsible management of water resources. In this effort, during the last few years, a wide variety of materials have been synthesized and tested for removing contaminants from water. In this work, we discuss the fundamental characteristics and properties of aerogels that have been used for the removal of organic and inorganic pollutants, including heavy metals, emerging contaminants and dyes. The raw materials, synthesis processes, characteristic structural features, and the main applications of silica, clay, carbon, graphene and polysaccharide-based aerogels are presented. These materials exhibit a large specific surface and have a tuneable porous structure that can make them highly effective adsorbents, which can also be easily removed from the media in which they are applied. Different types of aerogel composites are also discussed, emphasizing some of the advantages obtained by combining materials with different strengths. The adsorption capacity and removal efficiency of several aerogels used to remove pollutants by adsorption and

* Corresponding Author's Email: cristina.almeidan@epn.edu.ec.

In: Aerogels: Properties and Applications
Editor: Alina Iuliana Pruna
ISBN: 978-1-68507-788-4

photocatalytic degradation is presented. Finally, we provide an outlook to the research required to understand the behavior of aerogels and tailor their structure and behavior aiming at improving their performance. One of the objectives of the work envisioned is the formulation of materials with desirable characteristics such as high removal ability, environmentally friendly nature and processing, low-cost, and recyclability, among others.

Keywords: wastewater, water treatment, pollutant, emerging contaminant, heavy metal, dyes, adsorption, photocatalytic degradation

Introduction

Population growth, industrialization and urbanization have increased water consumption and wastewater generation worldwide. The type of pollutants and the amount of wastewater produced is related to the activities and lifestyle of the population [1]. Approximately 330 km^3 of municipal wastewater is produced annually, however only 60% receive some type of treatment. Nearly 2.4 billion people cannot adequately access sanitation facilities. This has caused the contamination of water resources and public health problems (6.3% of deaths worldwide is associated with contaminated water) [1, 2]. Additionally, phenomena such as climate change and the inadequate use of natural resources continuously reduce the availability of renewable water resources. This demands the development of strategies, plans and technologies for a responsible management of water resources.

Organic and inorganic compounds, heavy metals, and biological organisms are the main pollutants present in wastewater. However, a few decades ago it was determined the presence of a wide variety of substances (more than 1000) in water sources and the environment. These substances are present in very low concentrations (ng L^{-1} to μg L^{-1}) and are called emerging contaminants. The emerging contaminants come from product residues used in large quantities in daily life. Some examples of emerging contaminants are pharmaceuticals, personal care products, illicit drugs, food additives, pesticides, endocrine disruptors, among others [3, 4].

To reduce the problems associated with wastewater, pollutant removal processes are carried out. There are several processes, technologies and materials for the removal of pollutants. Their selection and use are associated with the quantity and quality of wastewater and the available budget [5].

Conventional centralized wastewater treatment plants generally consist of three stages: primary treatment (i.e., sedimentation), secondary treatment (i.e., activated sludge systems, coagulation - flocculation) and tertiary treatment (i.e., filtration, disinfection) [6]. Centralized treatment plants have proven to be efficient in removing organic matter (around 90%) and nutrients (up to 70%) [5]. However, the technologies used in conventional treatment plants are at least 50% more expensive than nonconventional technologies. Therefore, improving the pollutant removal in wastewater by developing low-cost, versatile, easy to operate, ecological and efficient alternatives has become a challenge worldwide [7].

Adsorption and photocatalytic degradation are techniques that are being extensively investigated for the removal of pollutants present in wastewater because they are considered efficient and economically scalable treatments [8]. In both of these techniques, the material used plays a very important role. The most widely used adsorbent is activated carbon since it has proven to be very efficient in the removal of several pollutants, this thanks to its high surface area (greater than 1500 $m^2 g^{-1}$), high porosity, and high degree of surface interactions. However, it has some disadvantages such as high cost, high operational costs (20-22 USD kg^{-1} and between 0.13-0.31 USD m^{-3} of treated water, respectively), difficult and costly recovery, and decreased adsorption efficiency after regeneration. Other organic and inorganic adsorbents have been investigated. However, these adsorbents have some disadvantages. In the case of inorganic adsorbents, they have limitations such as high density and low oil adsorption capacity. On the other hand, most of the organic adsorbents presented problems associated with their low hydrophobicity (they adsorb water) [6, 9–11].

In photocatalytic processes, the most used material is TiO_2, specifically Aeroxide® TiO_2 P25 due to its crystalline phases, particle size, high efficiency (up to 100%), among others. However, its recovery and regeneration are complicated. Its photocatalytic capacity can be reduced over time and its degradation efficiency decreases if the pollutant concentration is very high [10, 12, 13]. Currently there are investigations in which the adsorption and photocatalysis processes are combined to increase the efficiency of removal of contaminants [14].

To solve these problems, other alternative materials have been investigated for the pollutant removal from wastewater. These materials have advantages such as high adsorption capacity, selectivity, density, recyclability, and need to be hydrophobic, lipophilic, and also environmentally friendly. Among these are the aerogels. Aerogels have proven to be very efficient in removing several pollutants, due to characteristics such as high surface area, high porosity, presence of active sites, photocatalytic properties, among others [11].

Aerogels in the Removal of Pollutants Present in Wastewater

Aerogels are gels that replace the liquid present in its porous structure with gas. In their synthesis they can be given a specific chemical surface (designing the surfaces, doping them, etc.), varied shapes and sizes. For instance, in the sol-gel process depending on the growth process and the crosslinking rate, groups of networks will be obtained in the polymeric or colloidal morphology in the size range of 10-100 Å. In the aging process, however, the pore size, porosity and surface area can be modified. Aerogels are characterized by having unique properties as low density (0.003-0.5 g cm^{-3}), high surface area (500-1200 m^2 g^{-1}), high porosity (80-99.8%), low pore diameter, abundant three-dimensional (3D) porous structure and easy regeneration (more than five adsorption-desorption cycles). In addition, aerogels have excellent thermal, acoustic, electrical and water barrier capacity [15–18].

Aerogels combine the chemical and physical properties of nanomaterials with macroscale materials, integrating the specific properties of their constituents. This allows aerogels to be attractive for different applications (thermoresistors, piezoelectrics, and sensors), including the pollutant removal. In recent years, this application has been important due to the efficiency that aerogels have had in the removal of pollutants present in air (mainly CO_2) and wastewater (heavy metals, dyes, dangerous organic compounds, among others) with high selectivity [15, 19].

In addition, in recent years, aerogels have been studied for the oil removal from water. Removal efficiencies between 16 and 356 g g^{-1} have been achieved, using mechanical processes, extraction, distillation, combustion, and evaporation for oil recovery. The advantage of using these types of materials in oil and hydrocarbon spill treatment is that the pollutant can be recovered and the aerogel recycled [20].

Since their appearance, several types of aerogels have been investigated such as inorganic (i.e., metal oxides, carbide, silica-metal oxide and metal-doped, metal sulfides, selenides and tellurides), inorganic-organic hybrids (i.e., cross-linked polymeric inorganic aerogels), organics (i.e., resorcinol/formaldehyde, melamine/formaldehyde, polyurethane, polyamide, polyurea, polynorbornene, polydicyclopentadiene, polythiophene and polybenzoxazine) and carbon aerogels. Among the aerogels most used in the removal of pollutants are carbon, silica, polymer and naturally based aerogels [15, 19]. Figure 1 illustrates the main mechanisms and different types of aerogels that are used to remove pollutants from water.

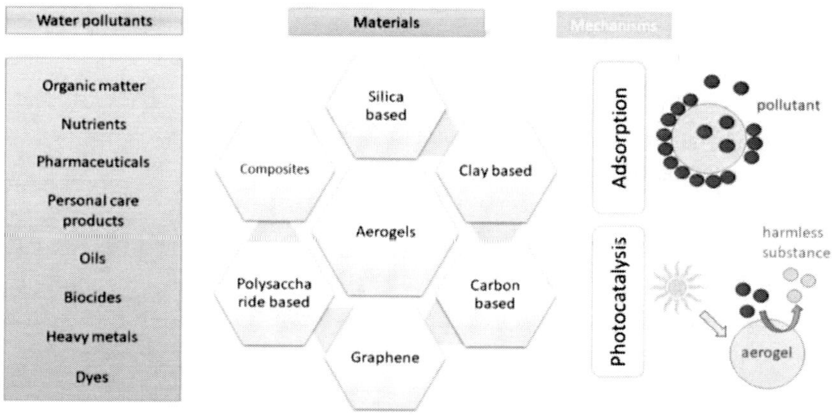

Figure 1. Mechanisms and types of aerogels used to remove pollutants from water.

Silica Based Aerogels

Silica aerogels are amorphous and mesoporous materials that can be synthesized by the sol-gel method. They are characterized by having a large surface area (500-1200 $m^2 g^{-1}$), narrow pore size distribution (pores diameter between 20 and 40 nm), high porosity (80-99.8%), flexible processability, low bulk density (0.003-0.5 $g cm^{-3}$) and low thermal conductivity [20, 21].

One of the advantages of silica-based aerogels is that their surface can be modified in such a way that it can remove several pollutants. The incorporation of organic groups in the silica network, using organosilane precursors (which add amino groups, thiol, acid, etc. to the silica molecule) has increased the efficiency in the removal of heavy metals, phenolic compounds, among others. Precursors such as 3-aminopropyl (trimethoxysilane), mercapto propyl trimethoxysilane have been used to incorporate amine groups in the silica aerogels. This resulted in an increase in the surface in a range between 240 to 920 $m^2 g^{-1}$. The modified aerogels were used in the removal of heavy metals such as Cr (III), Cd (II), Cu (II), Ni (II), Pb (II), Zn (II) and Hg (II) obtaining efficiencies higher than 90% [16, 22–24].

In addition, silica aerogels have been used in the separation of oil from wastewater. The aerogels have completely removed the oil (oil/aerogel ratio= 3.5), obtaining a dry solid when easily separated from the aqueous system. The silica aerogels have undergone chemical modifications to improve their properties. Among the substances used are CF_3 and $Si(CH_3O)_4$ [26].

Clay Based Aerogels

Clay aerogels have a large interlaminar spacing (1 to 4 nm), high porosity (between 50 and 80%), high specific surface (100 to 800 m^2 g^{-1}) and low density (around 0.8 g cm^{-3}). The properties depend on the parameters used in the synthesis process (variety and dose of the modifier, the gel acidity-basicity, among others). Clay aerogels are easily synthesized through an environmentally friendly process that makes them suitable for industrial production [27].

Montmorillonites are the most versatile type of clay; they are used to obtain honeycombs by extrusion (good radial crushing) since they do not present defects during drying. The ordered hydrophilic structure of the montmorillonites allows the synthesis of aerogels since they show a good and stable dispersion and the ability to exfoliate in the polymer used for the synthesis. When sodium montmorillonite is dispersed in water in concentrations greater than 2 w/w %, a thermodynamically stable gel is formed that exhibits a house of cards structure. The cooling temperature influences in the gel dimensions. Calcium montmorillonite does not form structures, but rather expanded powders. On the other hand, kaolinites are used in the synthesis of ceramic foams because they sinter at very high temperatures. Kaolinites do not form structures if they are in concentrations lower than 20 w/w % since they interact less strongly. Clay aerogels have been used in the removal of volatile organic compounds, colorants (methylene blue), heavy metals (lead, cadmium and chromium) [26–28].

Carbon Based Aerogels

The properties of carbon materials make them useful for several applications (energy storage materials, catalytic scaffolds, and adsorbents). However, the presence of impurities has promoted the synthesis of synthetic carbon materials such as carbon aerogels. Carbon aerogels can be obtained from the carbonization of organic aerogels, which are prepared from the sol-gel polycondensation of certain organic monomers, such as resorcinol, formaldehyde, among others. A carbon aerogel is a synthetic gel, formed from a three-dimensional network (interconnected nanometric primary particles). They can be in the form of monoliths, beads, powders or thin films. Carbon aerogels are characterized by being highly porous (it does not have structural contraction) with controlled micro- and mesoporosity, which gives it properties such as very low density (0.003 g cm^{-3}), low thermal conductivity, sound insulation and a large specific surface [15, 18, 29].

Among the carbon-based aerogels are carbon nanotube (CNT), carbon fiber, carbon microbelt, and graphene aerogels. They present 3D structures, low density, high porosity, large specific surface area and surface hydrophobicity. They have a sorption capacity of up to 192 times the aerogel weight. CNTs with porosity greater than 99% show remarkable selectivity and adsorption capacities up to 180 times their weight [15].

Precursors and equipment for the synthesis of graphene aerogels, CNT, microfibers can be expensive. Finding inexpensive, easy-to-use and environmentally friendly methods, as well as highly available materials for the synthesis of carbon aerogels, has become a challenge. Due to characteristics such as low cost, high availability, natural porous structures, high surface area and good sorption capacity, agro-industrial residue is presented as an alternative. Among the residues used for the synthesis of aerogels that are applied for the removal of pollutants present in wastewater are pomelo peels, sago pith waste, cotton waste, durian shells, watermelon, winter melon, sugarcane bagasse, etc. [30, 31]. Zhu et al., 2017, prepared carbon aerogels from pomelo peels obtaining an aerogel with a specific surface between 466.0-759.7 m^2 g^{-1}, a pore size between 2.38-2.60 nm and a pore volume between 0.30 and 0.45 cm^3 g^{-1}. Pomelo peel aerogel was used in the adsorption of ethanol, dimethylformamide, acetone, p-xylene, n-heptane, toluene, methanol, formic acid, 1- methyl-2-pyrrolidone [32].

Graphene Aerogels

Graphene aerogels (GA) have ultralow density (around 5 mg cm^{-3}), higher compressibility, higher specific surface area (around 800 m^2 g^{-1}) and porosity (up to 500 cm^3 g^{-1}), and better electrical conductivity than other graphene macrostructures. The characteristics of graphene aerogels added to their easy interaction with several substances, high absorption capacities, rapid absorption rates, chemical inertness, stability in acidic/basic media, have made possible its use in the adsorption of pollutants present in wastewater [33].

Graphene oxide and graphene are considered efficient adsorbents to remove pollutants from wastewater. However, due to their small size, excellent chemical stability and dispersibility in water, it is difficult to separate them from the aqueous medium and could form secondary contamination. To solve these problems, 3D aerogels have been synthesized [34].

GA form a 3D network with an interconnected microporous, mesoporous and macroporous structure that increases the surface area of graphene. This porous structure, combined with the functional groups found on the surface (carboxyl, hydroxyl and epoxy groups) make these aerogels good candidates for the removal of several pollutants by adsorption or photocatalysis processes. In addition, the porosity that these materials show generates an internally active surface that favors the adsorption process [33, 34].

3D GA have been used efficiently in photocatalysis, reaching up to 100% dye removal. 3D GA have favorable conductivity and a multidimensional electron transmission path, promoting the separation of photogenerated electron pairs and holes. In addition, the porous structure and the several functional groups present on their surface allow this type of aerogels to inhibit the aggregation and excessive growth of semiconductors, showing more active sites for surface catalytic reactions to occur [36]. 3D GA also have a high adsorption capacity for organic pollutants, thus increasing their ability to degrade them. In adsorption processes they have been efficient for the removal of heavy metals, dyes, oils, etc. Dye adsorption was carried out through π-π conjugation, electrostatic interaction, and hydrogen bonding [33, 35].

To increase the adsorption capacity, the adsorption rate and facilitate the regeneration of graphene 3D aerogels, they have been combined with other materials (MnO_2, carbon nanotubes, lignosulfonate, among others) to form composites [35].

Polysaccharide-Based Aerogels

Polysaccharides are carbohydrate polymers formed by a set of monosaccharides (homo/hetero repeats) connected by glycosidic bonds. Polysaccharides come from plants and as by-products of industrial processes, they have structures that range from linear to highly branched [37]. The development of polysaccharide aerogels has been stimulated by the fact that conventional aerogels (silica, graphene, titanium, or their oxides) come from inorganic materials or from petroleum derivatives. Among the precursors of polysaccharide aerogels are cellulose, hemicellulose, marine polysaccharides, starch, chitosan, alginate, carrageenan, and pectin. These materials are characterized by being biocompatible, biodegradable and have low toxicity and high renewability. The aerogels obtained have the same characteristics, added to the high porosity (up to 99%), relatively large surface area (31–479 $cm^2 g^{-1}$) and low density (0.02-0.17 g cm^{-3}) [38]. Some characteristics of some polysaccharide aerogels and their application in the pollutant removal are described below.

Cellulose Aerogels

Cellulose is the most abundant biopolymer, it is a water-insoluble homo-polysaccharide, consisting of linear chains of connected anhydro-glucose units. Unlike synthetic aerogels, in cellulose-based aerogels the starting material is a polymer that interconnects three-dimensionally to form a sponge. Cellulose aerogels cannot be synthesized by the sol-gel method since cellulose structures collapse in protic media and with traditional drying. To improve the gelation of cellulose aerogels, various substances have been incorporated such as graphene, silanols, ethanol, among others [17].

Cellulose aerogels have low density (between 0.01 and 0.1 g cm^{-3}), high specific surface area (up to 600 g m^{-2}) and porosity (between 41.0 and 99.7%). The specific surface area of the cellulose aerogels is a function of the drying method adopted, and the cellulose fibril width used, while the density and porosity depend on the nanocellulose initial concentration. Furthermore, the cellulose aerogels are very flexible with good mechanical stability [17, 37, 38].

Cellulose aerogels have an advantage over other aerogels, since they can be synthesized from residues (e.g., paper, corn straw, wheat husk, rice husk, etc.), they are biocompatible, biodegradable, and hydrophilic. Furthermore, they turn out to be very interesting due to their nanosized structure, favorable to charge and release molecular hosts [17, 38]. Due to their characteristics, cellulose aerogels can be used in environmental applications, among them the removal of pollutants in wastewater. Due to their wettability, they can separate hydrocarbons, oils and lubricants from water (reaching efficiencies greater than 99%), with modifications in their surface they have been efficient for the removal of contaminants (i.e., heavy metals, with adsorption capacities of up to 700 mg g^{-1}) [36, 39].

Chitosan Aerogels

Chitosan is an abundantly available low-cost biopolymer that comes from chitin (second most abundant polysaccharide). Chitosan is characterized by being non-toxic, heterogeneous, linear, cationic, insoluble (water, basic solutions, and organic solvents), biocompatible and biodegradable. It also has a high molecular weight (which influences the removal of contaminants). Based on these properties and the diversity of functional groups on its surface (amine and hydroxyl group), chitosan has shown to have a high adsorption capacity for several pollutants (i.e., heavy metals, dyes). However, its instability in an acid medium, its low mechanical resistance and its low surface area have limited its efficiency in adsorption processes. This has promoted the investigation of possible modifications such as the aerogel synthesis [40, 41].

The presence of functional groups in the structure of chitosan facilitates its dissolution, being able to crosslink to form the three-dimensional network. The porous chitosan aerogel obtained by extracting the liquid solvent in the hydrogel has a higher adsorption capacity, an easier and more environmentally friendly separation process. In addition, chitosan aerogels have high porosity (higher than 90%), low density (between 0.0283 to 0.92 g cm^{-3}), high specific surface (up to 845 m^2 g^{-1}), low coefficient of thermal conductivity and low dielectric constant [44]. Chitosan aerogels have been used in the removal of heavy metals, dyes, and organic compounds.

To improve the properties of chitosan aerogels cross-linkers (glutaraldehyde, glyoxal, formaldehyde, konjac glucomannan) are added. Among the properties that are improved are adsorption capacity due to the introduction of functional group with coordination atom, chemical stability and mechanical properties. Another option is the composite formation, in its synthesis is also recommended the use of cross-likers [40, 43].

Starch Aerogels

Starch is a highly available and inexpensive polysaccharide; it occurs naturally in the form of semi-crystalline granules formed from amylose and amylopectin. The alternating amorphous and crystalline layers located within the starch granules result in different crystallization types. In a natural way, starch has some spaces (low specific surface and pore volume) for the adsorption and encapsulation of external substances. Therefore, it is necessary to modify the natural starch to increase its porosity and its specific surface area [46].

Porous starch is a low-cost, non-toxic adsorbent with applications in the food, pharmaceutical and environmental industries. To improve their properties, the starch aerogels have been carbonized, reaching specific surfaces greater than 1500 m^2 g^{-1} [47]. Starch aerogels have been used in the removal of heavy metals (i.e., Zr (IV), Al (III), Fe (III), La (III), Pb (II) and Cu (II)) and dyes (i.e., methylene blue, crystal violet, methyl violet) [45,46]. To functionalize the porous starch, it has been mixed with other polysaccharides (i.e., pectin and chitosan) and inorganic materials (i.e., graphene oxide and $CoFe_2O_4$). In addition, composites with graphene oxide have been synthesized [46].

Titanium Dioxide Based Aerogels

TiO_2 has been widely used in photocatalytic degradation processes because it is photo-oxidant, has good chemical and thermal stability, is non-toxic, shows resistance to photocorrosion, has low cost and is friendly to the environment. It is also highly available (Ti represents 0.44% of Earth's crust). However, TiO_2-based photocatalysis occurs only with UV light ($\lambda < 380$ nm) due to the wide band gap (3.2 eV) and the rapid recombination of photogenerated TiO_2 electron and hole pairs, which limits its practical application. Combining other materials with TiO_2 could allow the photocatalysis with visible light irradiation in a way that improves the efficiency of TiO_2 [48].

To increase the photocatalytic yield of TiO_2, several synthesis methods have been developed, in which different sizes, porosity, morphology and controlled composition have been obtained. The photocatalytic activity of TiO_2 could be increased by increasing the useful life of the holes and by reducing the rate of recombination of electrons and holes to produce OH• radicals. TiO_2 aerogels are the most widely used photocatalysts in the form of aerogel, they have become an option for photocatalysis processes due to their high porosity, high surface area (between 65 and more than 1000 m^2 g^{-1}), interconnected pores and crystallinity. TiO_2 has been used in the degradation of emerging contaminants, dyes (i.e., methylene blue, rhodamine B, methyl orange, indigo carmine), other organic pollutants, and heavy metal ions, reaching removal efficiencies greater than 50% [45–48].

Metal-Organic Framework-Based Aerogels

Metalorganic structures (MOF) consist of organic bridging ligands and metal ions. They are crystalline porous materials, which have received great attention for their characteristics [53]. MOF aerogels have high surface areas (of the order of 1000 $m^2 g^{-1}$), high microporosity (2 nm < diameter < 50 nm, 0.5 - 0.7 $cm^3 g^{-1}$), high thermal stability and tunable pore structure. MOFs have been used in some applications such as gas separation and storage, chemical detection, catalysis, etc., proving to be very efficient [20, 48].

To improve some properties of the MOFs, they have been carbonized. The carbonized MOFs increased their specific surface area due to their layer separation. They reached values between 400 and higher than 1700 $m^2 g^{-1}$ and have been used in the removal of pollutants such as methylene blue, emerging contaminants (i.e., sulfamethoxazole, bisphenol A, diclofenac, ibuprofen, atenolol, clofibric acid), and copper ions, achieving adsorption capacities greater than 300 mg g^{-1}. Carbonized MOFs have been modified using chemicals such as 1-butyl-3-methylimidazolium bromide to modify the surface of the aerogel (increase the nitrogen in the MOF surface) and KOH, KCl, LiCl, NaCl to increase the hierarchical structures, porosities, and performance of several carbon materials. Emerging contaminants such as atrazine, herbicides (diuron, 2,4-dichlorophenoxyacetic) were removed [54].

Noble Metal Aerogels

Noble metals are rare in nature and highly resistant to oxidation; among these are gold, silver, ruthenium, rhodium, palladium, osmium, iridium, and platinum. They are characterized by being expensive, so studies are carried out in an attempt to improve their performance/price ratio. Noble metal aerogels have succeeded in combining the physicochemical properties of noble metals and the structural attributes of foams. That is, they have numerous active sites and abundant mass/electron transfer pathways, which is why they are widely used in catalysis (i.e., Pt, Ru, Rh, Pd, Ir, Au). These types of aerogels have various shapes (nanowires, nanotubes, nanofilts, nanocubes), ligament sizes between 3.1 and 142.0 nm, surface areas between <3 to 1300 $m^2 g^{-1}$ and ultra-low densities (up to 3.8 mg cm^{-3}) [29, 49].

Noble metals have been deposited on metal oxides such as γ-Al_2O_3, SiO_2, TiO_2, ZrO_2, Fe_2O_3, CeO_2, MnOx, among others. These aerogels have been used in the removal of salicylic acid and volatile organic compounds [56]. In the salicylic acid removal, a gold doped TiO_2 aerogel was used; the gold presence improved the photocatalytic activity of TiO_2 due to the formation of an efficient electron trap and improved the response of the photocatalyst to illumination. Furthermore, the pollutant diffusion is improved, and the number of metal/pollutant contacts is increased. Salicylic acid removal improved by up to 22% over TiO_2 alone [57].

This type of aerogels has been studied very little in the removal of pollutants from wastewater. However, in gas phase heterogeneous catalysis it has proven to be very efficient due to its very favorable characteristics. These include a high specific surface area, ease of mass transfer through pores, low diffusion resistance, and easy access to active sites (defects, facets, and left stabilizing species). In addition, aerogel networks are mainly composed of interconnected nanochains, they are in the optimal range for electrocatalysis, the nanochains interconnected in the networks and their self-supporting character can provide them with greater durability in electro-catalysis or catalysis processes. Finally, the molar ratio of the constituent elements in multimetallic aerogels can be precisely controlled, allowing the optimal balance between composition, reactivity, durability and cost to be designed and taking full advantage of synergistic effects to enhance catalytic activity. Due to these characteristics, its application in wastewater treatment is interesting, as long as the cost reduction of aerogels is defined.

Aerogel Composites

Composites are formed to improve or take advantage of some of the properties of the materials used. For instance, the combination of cellulose with MOFs improves and facilitates the MOF processing. The compression of the graphene sheets due to the π - π effect is improved with the addition of another material because the surface wettability and compressive strength are favored [20, 52]. Among the most widely used aerogel composites are the composites that form MOFs (MOFs-silica, MOF-graphene, MOF-cellulose, 3D TiO_2-reduced graphene oxide aerogel), silica-alumina, noble metals, graphene and cellulose composites [20, 21, 53].

MOF Composites

MOFs have been used in the synthesis of different composites to improve some of their deficiencies, such as gelling on crystallization, intrinsic rigidity and dispersion. MOFs have been combined with materials such as zirconium, graphene and its oxide, among others [54].

MOF composites have been used in the adsorption of pollutants present in wastewater. Among the adsorbed pollutants are heavy metals (i.e., Cr (VI), Pb (II), Cu (II)), dyes (i.e., Rhodamine B, methyl orange, crystal violet) and other organic compounds (i.e., hydroquinone, ciprofloxaxin, tetracycline hydrochloride) [21].

To synthesize more environmentally friendly aerogels, polysaccharides (agar, cellulose and alginate) have been used to obtain composites with MOF. Other polysaccharides have been used in the synthesis of MOF aerogels; their incorporation has improved the characteristics of cellulose (flexibility, physical stability, adsorption capacity) and MOF. The presence of cellulose allows obtaining a fibrous porous matrix combined with MOF [54].

MOFs-cellulose composites have been used to remove Cr (VI), reaching removals greater than 90%. Composites such as MOFs-cellulose (ZIF9-cellulose and ZIF12-cellulose) have been used in the degradation of organic pollutants such as P-nitrophenol, Rhodamine B and dimethyl phthalate, reaching a degradation of over 80%, 100% and 85%, respectively. For the Cr (VI) removal, UiO-66@cellulose aerogel (removal between 67 and 85%) and ZIF-8-cellulose aerogel (removal between 90.0 and 99.7%) have been used [20, 48].

Carbon aerogels derived from MOFs have also been efficiently used in the removal of organic pollutants such as n-hexane, and take up around 60 times their weight [54]. Some other examples of these composites and their efficiencies in removing a series of pollutants are presented in Table 1.

Silica-Alumina Composites

Silica and alumina aerogels have compact particles interconnected with a spongy structure. Aerogels obtained from synthetic precursors reach surface areas between 278 and 907 $m^2\ g^{-1}$, bulk densities around 0.57 $g\ cm^{-3}$, large pore sizes (11.4-36.4 nm) and pore volumes greater than 5.0 $cm^3\ g^{-1}$. The specific surface decreases while the pore size increases with increasing Al content. The properties of the silica-alumina aerogels depend on their composition. It was determined that the aerogels with the highest Al content have low bulk densities (0.05-0.41 cm^{-3}), with a pore size between 31.7 nm-100.0 μm and high specific surface area (250-780 $m^2\ g^{-1}$).

An advantage of silica-alumina aerogels is that they can be synthesized from inexpensive materials, such as waste (kaolin, palygorskite, coal gangue, and fly ash). The aerogels obtained show large specific surface areas (136-930 $m^2\ g^{-1}$), low bulk density (0.09-0.45 g cm^{-3}) which is even less than that of the aerogels obtained by synthetic precursors, pores with sizes between 0.5 and 11.0 nm and a pore volume between 0.59 and 2.7 $cm^3\ g^{-1}$.

The aerogels of the silica-alumina composites have been used in the adsorption of metals and organic substances. These adsorbents have achieved high removal efficiencies (Table 1) due to the presence of OH^- groups in their inner and outer surface, high surface area and porosity [21, 23, 25, 54, 55].

Graphene Composites

The unique properties of 3D GAs (adsorption capacity and unique monolayer structure) can be combined with the properties of other materials (i.e., semiconductors, metal oxide semiconductors, metallic and/or nonmetallic ions, carbon materials) to form composites. Another alternative is to combine graphene oxide (a precursor to 3D graphene aerogels) with soluble precursors of photoactive materials such as metal salts. The composites formed show very low density (between 3 and 54 mg cm^{-3}), high surface area (higher than 400 $m^2\ g^{-1}$) and ample pore volume (higher than 2 $cm^3\ g^{-1}$) [35, 56].

In the synthesis of these composites, properties such as crystallinity, capture state and recombination sites (due to the presence of fewer defects), adsorption capacity, a wider light absorption range, a separation of faster charging and mass transfer, have improved their degradation capacity. Among the composites used in the degradation of pollutants in wastewater are 3D GA-TiO_2, 3D GA-BiOBr, 3D GA-Bi_2MoO_6, 3D GA- g-C_3N_4, C_3N_4-graphene oxide aerogel, Cu_2O-g-C_3N_4-GA, MOF-GA, 3D GA-Bi_2WO, carbon nanotubes, among others. These composites have been used in the degradation of organic matter and colorants (i.e., rhodamine B, methylene blue, methyl orange), reaching degradation efficiencies greater than 50% [36].

Nanocellulose-Based Composites

Nanocellulose is a natural polymeric material that has good biocompatibility, degradability, and a wide range of sources. It is characterized by being hydrophilic and superoleophobic under water. Nanocellulose-based aerogels are friendly to the environment since they are derived from renewable resources and their processing does not require harmful solvents. The structure of the porous nanocellulose material collapses on contact with water so compositing can solve this problem. Nanocellulose aerogels in their different forms (nanofibers, nanocrystals, cellulose, and bacterial cellulose) are being used in different applications in the cosmetic, food, water treatment, medical and energy industries. Nanocellulose-based aerogels can be used as solid support for nanoparticles, which facilitates their separation after adsorption and photocatalysis processes [37, 57, 58].

Among the composites used are mixtures with chitosan, graphene oxide, graphene, gold nanoparticles, silver nanoparticles, TiO_2, MOFs, polyvinyl alcohol, among others. The cellulose composites presented a porosity greater than 97% and low density (0.2-0.4 mg cm^{-3}). Nanocellulose composites have been used in the removal of pollutants (adsorption and photocatalysis) such as oils (i.e., paraffin oil, diesel, motor oil), heavy metals, organic pollutants (i.e., 2-naphthol, linuron, nitrobenzene, phenol, xylene, atrazine, among others) and dyes (i.e., Victoria blue, methyl violet, rhodamine 6G) [37, 57, 59, 60]. Cross-linking agents (isocyanate resin, carbodiimide and polyamide epichlorohydrin) are used to improve the properties of nanocellulose aerogels [39].

TiO_2 Composites

To improve the photocatalysis of TiO_2, it has been mixed with other materials such as metal ions, carbon, graphene and other semiconductors. For instance, the carbon presence can change the photocatalytic behavior of TiO_2, which goes beyond the synergistic effect on degradation kinetics. On the other hand, the immobilization of the TiO_2 mesopore on porous carbon foam (CF) has shown a higher photocatalysis activity compared to TiO_2 powder. The synthesized TiO_2 composites include WO_3-TiO_2, Au-TiO_2, titanate nanotube-TiO_2, polyaniline-TiO_2, SiO_2-TiO_2, syndiotactic polystyrene-TiO_2, Pt-TiO_2, TiO_2-Au, graphene oxide-TiO_2, among others [45, 61, 62].

The composites of TiO_2 aerogels have reached high surface areas (140-572 m^2 g^{-1}). They have been used to remove emerging contaminants (i.e., caffeine, ibuprofen, carbamazepine, sulfamethazole) and dyes (i.e., crystal violet, methylene blue, congo red, methyl orange) and degradations of up to 99% have been achieved [62, 63].

Polymeric Composites

Polymers are good candidates to be synthesized as aerogels because they are lightweight, easy to process, corrosion resistant, chemically stable, and exhibit abundant functional groups on the surface that can be easily modified with a variety of materials [70]. To improve the physical properties of polymers, inorganic compounds that can have different forms (platelets, spheres, fibers, etc.) are dispersed in them. Among the composites formed are polyaniline-montmorillonite, polyaniline-V_2O_5, layered double hydroxide of polyvinyl alcohol, poly-N-vinyl carbazole-graphene oxide, polyvinyl alcohol-graphene, among others.

Mesoporous silicas have been used as supports for polymer preparations. In addition, to improve their properties or to win photocatalityc properties, polymers have been doped with nanoparticles such as TiO_2, Au, CdSe, γ-$[SiW_{10}O_{36}]^{8-}$. Thus, polymer aerogels are used in both adsorption and photodegradation processes [71].

Polymer aerogels have reached specific surface areas of up to 800 m^2 g^{-1}, porosities greater than 90% and low density (0.16-0.24 mg cm^{-3}) [65, 66]. Polymer aerogels have been used in the removal of oils, since they can be compressed and recycled under a certain pressure to discharge the adsorbed oil, this simplifies the separation process [70]. In addition, polymeric aerogels are used in the heavy metal removal (i.e., copper, lead, cadmium, nickel, zinc, iron (III), cobalt, gold (III), chromium (VI), mercury), reaching adsorption capacities between 34.9 and greater than 1000 mg g^{-1}. Organic compounds (i.e., methanol, non-steroidal anti-inflammatory drugs), dyes (methylene blue, Rhodamine B) and oils are also removed [69–71].

Polymeric aerogels are currently being synthesized from recycled materials such as polyethylene (high and low density), polyethylene terephthalate, and polypropylene. They have been used in the removal of oils, reaching removal efficiencies of up to 100 mg g^{-1} [74].

Aerogels Used in Adsorption Processes

The adsorption process is an efficient method to remove pollutants. Its advantages include simplicity, low costs and high adaptability compared to the other technologies used [75].

Adsorption is a process in which a material (adsorbate) in a different state (solid, liquid or gas) or size (atoms, molecules, ions) is retained on a surface (adsorbent). In this case, the pollutant (adsorbate) is deposited on the adsorbent surface through electrostatic forces or chemical bonds. Pollutants in the wastewater are gradually adsorbed on the adsorbent surface until the adsorbent-adsorbate equilibrium is reached. To efficiently remove of pollutants from wastewater, adsorbents must have high surface area, porosity, adsorption and regeneration capacity, abrasion resistance, chemical and thermal stability, and be environmentally friendly [8]. In addition, since most adsorption processes are reversible, adsorbents can be regenerated by desorption processes, which allows the adsorbent to be used multiple times. Several desorption processes have low maintenance costs, high efficiency and are easy to operate [73]. Table 1 shows some aerogels used in the adsorption of several pollutants, the conditions for the tests carried out, and the adsorption efficiency or adsorption capacity of the aerogels.

Despite the advantages and the efficient adsorptive removal of organic pollutants using aerogels, adsorption has a disadvantage since the adsorbed pollutants cannot be degraded to less toxic substances and this can cause secondary pollution. Considering this, it is very important to develop materials with high adsorption and degradation capacity to remove these pollutants from the water/wastewater. Synergistic adsorption/degradation are efficient processes for the removal of pollutants present in wastewater.

Nanomaterials such as TiO_6, ZnO, CdS, Bi_2MoO_6, Ag_3PO_4, AgCl, AgBr and AgI have been used as photocatalysts. They are characterized by being highly available, having low cost and high stability and excellent optical and electrical properties. Even so, serious nanoparticle agglomeration, wide bandgap, high exciton binding energy, and high electron hole recombination rate, seriously hamper their applications. Aerogels with photocatalytic properties have solved some of these disadvantages [53].

Table 1. Aerogels used in adsorption of some pollutants

Aerogel	Pollutant	Adsorption conditions	Adsorption capacity	Ref.
Graphene oxide	Copper	PC= 50-75; pH= 6.3; T= 283 -313; CT= 0.5	AC= 17.7 -29.6	[33]
Anionic polyacrylamide/ GO aerogel	Basic fuchsin	PC= 400; CT= 72; AD= 0.5; T= 293	AC= 1034.3	[34]
Thiol-functionalised cellulose sponge	Hg (II)	PC= 100; pH= 1 – 6; T= 298; AD= 0.5	AC= up to 700	[40]
Cross-linked chitosan-based aerogels	Sodium dodecylbenzene sulfonate	PC= 50; T= 278; pH= 2-12 CT= 24; AD= 0.2	AC= 869 - 1800	[42]
Chitosan aerogel monolith	Copper	PC= 30; T= 278; CT= 12; AD= 0.2	AC= 35.1	[44]
Starch-derived carbon aerogels	Crystal violet	PC= 500; pH= 2 – 10; T= 298; CT= 24 AD= 0.2	AC= 1515	[47]
	Methyl violet		AC= 1423	
	Methylene blue		AC= 1181	
gC_3N_4-TiO_2-graphene aerogel	Rhodamine B	PC= 20.0; AD= 0.2; CT= 1	96.5	[48]
Carboxymethyl cellulose–graphene oxide aerogel	Rhodamine B	PC= 50; CT= 24; AD= 0.002 – 0.01	AC= 186.3	[58]
Hydrophilic silica aerogel	Methylene Blue	PC=12.5 – 100; pH= ~7; CT= 72 AD= 20	AE= 97.2-99.5	[60]
	Malachite Green		AE= above 98	
	Gentian Violet		AE= above 98	
Silica-alumina composite	Pb	PC= 1000; CT= 1; AD= 0.5 – 3.0 mL	AE= 96.3 – 100.0	[61]
Gelatin/TiO_2/ polyethyleneimine	Methylene blue	PC= 20 – 500; CT= 10; AD= 6; pH= 5-9	AC= 40.1	[69]
	Congo red		AC= 86.7	
	Crystal violet		AC= 64.3	
	Methyl orange		AC= 85.2	
	Cu	PC= 10 – 500; CT= 24; AD= 0.5; pH= 4.6	AC= 61.2	
Polydopamine modified-chitosan aerogel	Cr(VI)	PC= 50-600; pH= 2-8; CT= 24; AD= 1	AC= 374.5	[76]
	Pb(II)		AC= 441.2	
Amine grafted pumice-derived silica aerogel	Ibuprofen	PC=6.53; pH= ~ 7; CT= 2.5; AD= 0.5	AE= 100 AC= 39.95	[77]

Table 1. (Continued)

Aerogel	Pollutant	Adsorption conditions	Adsorption capacity	Ref.
3D reduced graphene oxide/TiO_2 aerogel	Carbamazepine	PC= 10; CT= 0.67; AD= 0.5	AC= 3.9 - 12.5 AE= 55	[78]
Reduced graphene oxide–TiO_2/sodium alginate 3-dimensional aerogel	Sulfamethoxazole	PC= 10; CT= 0.75; AD= 0.5	AE= 50	[79]
	Ibuprofen	PC= 10; CT= 1.5; AD= 0.5	AE= 20	
Zeolitic imidazolate framework -Magnetic reduced graphene oxide	Malachite green	PC= 600; CT= 2; AD= $2.5*10^{-5}$	AC= 1850.0	[80]
Magnetite-reduced graphene oxides aerogels	As	PC= 0.1-10; CT= 24; AD= 0.5; T= 298	AC= up to 11.3	[81]
Silica-alumina composite	Ni	PC= 50; pH= 2 – 7; T= 283 – 318; CT= 0.5; AD= 1	AC= 49.8	[82]
Polybenzoxazine aerogel	Cd	PC= 1; pH= 4 – 5; CT= 4 - 48 AD= 0.2-1.6; T= 298	AE= ~20	[83]
	Cr		AE= ~20	
	Fe		AE= ~30	
	Cu		AE= 20-60	
	Ni		AE= 5-20	
	Pb		AE= 20-30	
	Sn		AE= ~100	
Poly(amidoamine)-grafted cellulose nanofibril aerogels	Cr (VI)	PC= 10 – 600; pH= 1-10; CT= 24	AC= 377.36	[84]
ZIF-8-cellulose nanofiber-cellulose sponge	Rhodamine B	PC= 666.7	AC= 25	[85]
	Cr (VI)	PC= 824.8	AC= 35	
	Dimethyl formamide	PC= 1000	AC= 45	

Pollutant concentration (PC, mg L^{-1}), Contact time (CT, h), Temperature (T, K), Aerogel dose (AD, g L^{-1}), Adsorption capacity (AC, mg g^{-1}), Adsorption efficiency (AE, %).

Aerogels Used in Photocatalysis

Photocatalytic degradation is a type of advanced oxidation process (AOP) that is used to remove organic pollutants (recalcitrant, dangerous, non-biodegradable, in low concentration "traces") and microbial pathogens from wastewater. AOPs are of great interest in water treatment, due to their high efficiency, gentle pH value of solution and mild reaction temperature. Photocatalysis is powered by solar energy and does not require a high operating temperature, making it attractive for the removal of pollutants [14, 33, 35].

In the photocatalytic process, organic pollutants are broken down into intermediates or CO_2 and water, thanks to the $\cdot OH$, $SO_4^{\cdot -}$, $^1O_2$ and $\cdot O_2^-$ that are produced in the photocatalytic processes, which attack the pollutants and break them down. This process is carried out during the interaction of the photocatalyst, a light source and an oxidizing agent (oxygen or air) [36].

The photocatalytic degradation of a pollutant follows the following stages: (i) Pollutant transfer from the aqueous system to the photocatalyst surface, parameters such as surface area, porosity and surface charges of the photocatalyst can influence the process; (ii) the pollutant is adsorbed on the catalyst surface with photon excited reaction sites; (iii) redox reactions of the photon activated sites occur with the adsorbed pollutant; (iv) part of the generated by-products and the resulting mineralization products (CO_2 and H_2O) are desorbed from the surface of the catalysts and leave the active sites free for subsequent reactions; and (v) the desorbed by-products are transferred from the catalyst interface to the aqueous medium. By-products that were not degraded repeat the process again until they are completely degraded [49].

Powder photocatalysts from semiconductors (TiO_2, ZnO, SnO_2, WO_3, Fe_2O_3 and CdS) have been used in the last decades. These materials have shown to be very efficient in the degradation of several organic pollutants due to their high photocatalytic efficiency, a favorable combination of their electronic structure, light absorption properties, easy synthesis, etc. However, at the nanoscale, the collection and separation of powder photocatalysts from an aqueous medium is complex, expensive and slow (it is not suitable to use in large scale industries) [33, 80].

The aerogels are catalogued as good photocatalysts because of characteristics such as high surface area, integrity of the structure, mechanical robustness, high porosity, easy separation and no major changes in their physical structure even after several reaction cycles. A photocatalytic aerogel can be obtained using a metal alkoxide precursor with an appropriate photocatalytic activity. Aerogels used as photocatalysts could be inorganic (i.e., SiO_2, TiO_2, ZnO, $InVO_4$), organic (i.e., graphene, boron, polyaniline-decorated cellulose) or hybrid (i.e., Cu_2O-reduced graphene, TiO_2-resorcinol-formaldehyde polymer aerogels composites, TiO_2-carbon aerogel). In addition, multi-function composite aerogels (exhibiting magnetism, catalytic characteristics or special wettability) are synthetized. They are typically manufactured using sophisticated, long-lasting, high-cost processes [69].

However, silica and titania aerogels are mostly used because they are non-toxic, highly photostable, and inexpensive [44, 80].

In addition to these aerogels, modified or composite aerogels have been used for pollutant degradation. Hansapour et al., 2020, in their review indicate the use of nitrogen (N)-doped graphene aerogels tetraethylenepentamine, Bi_2MoO_6-reduced graphene oxide aerogel, nitrogen and sulfur codoped graphene aerogels, Cu doped Cu_2O-reduced graphene, oxide-cellulose hybrid aerogel, BiOI-Silica aerogel, CuS-cellulose based aerogel, Fe_3O_4@TiO_2-SiO aerogel, among others. The dyes removed were methylene blue, acridine orange, methyl orange, rhodamine. The experiments were carried out under UV light, solar irradiation but mostly under visible light, reaching removal percentages between 39 and 99% [52].

Factors such as the amount of the photocatalyst and organic pollutant, pH (polluting solution and the catalyst solution), the calcination temperature of the aerogel, the presence of scavengers (e.g., ammonium oxalate, isopropanol, N_2), and dopant content, can affect the photocatalyst efficiency [52]. Table 2 shows the conditions set in some research in which different aerogels were used in photocatalytic degradation of some pollutants present in wastewater.

In Figure 2, we show the possible mechanisms (adsorption and photocatalytic degradation) when aerogels are used for the removal of different types of pollutants present in wastewater.

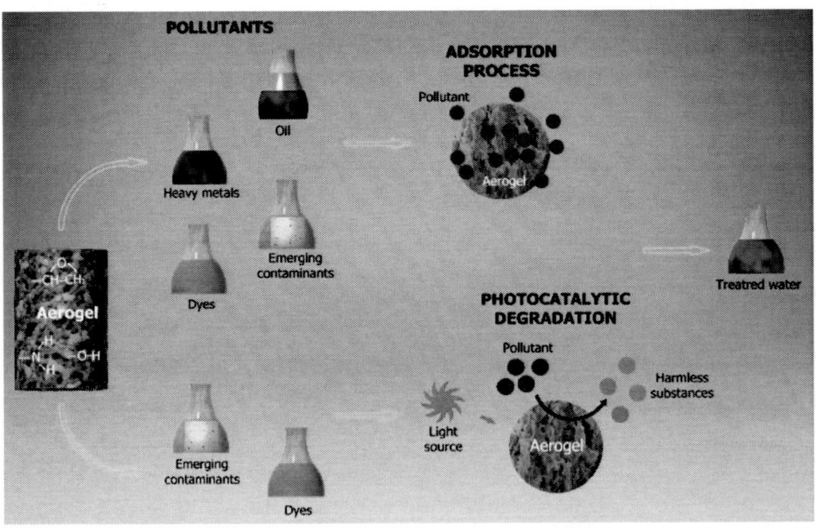

Figure 2. Mechanisms used to remove pollutants from wastewater using aerogels.

Table 2. Aerogel used in photocatalytic degradation of some pollutants

Aerogel	Pollutant	Photocatalytic conditions	Degradation (%)	Ref.
gC_3N_4-TiO_2-graphene aerogel	Rhodamine B	I= visible light; PC= 20.0; CT= 0.17; AD= 0.2	98.4	[48]
Mesoporous TiO_2 aerogel decorated onto 3D carbon foam	Rhodamine B	I=Solar-light; PC= 4.8; CT= 0.5; AD= 0.01	100.0	[50]
Nanocellulose-based aerogels decorated with silver nanoparticles	Methylene blue	I= UV light; PC= 10; CT= 0.08; AD= 0.02 T= 300	99.2	[63]
	Congo red	I= UV light; PC =10; CT= 0.35; AD= 0.02 T= 300	96.4	
Gelatin/TiO_2/polyethyleneimine	Methyl orange	I= UV light; PC= 5; AD= 8	-	[69]
SnO_2 aerogel@polymer-derived carbon (PDC)	Methylene blue	I= UV light; PC= 3.2; CT= 0.5; AD= 1.0	~90.0	[72]
3D reduced graphene oxide-TiO_2 aerogel	Carbamazepine	I= UV light; PC= 10; AD= 0.5	36.0-57.0	[78]
Reduced graphene oxide–TiO_2-sodium alginate 3D structure aerogel	Sulfamethoxazole	I= UV-A light; PC= 10; CT= 0.75; AD= 0.5	99.0	[79]
	Ibuprofen	I= UV-A light; PC= 10; CT= 1.5; AD= 0.5	78.0	
TiO_2	Methylene blue and rhodamine B	I= UV light; PC= 10.0; CT= up to 2; AD= 1.8	-	[86]
ZnS-graphene aerogel	Methylene blue	I= Sunlight; PC= 20; CT= 1	100.0	[87]
	Cr(VI)		61.2	
ZnO-graphene aerogel	Methyl orange	I= UV light; PC= 5; CT= 3 h; AD= 6 -15	100.0	[88]
BiOBr-graphene oxide aerogel	Rhodamine B	I= UV; PC= 20; CT= 3; AD= 60	100.0	[89]
Ag_2O-sodium alginate-reduced graphene oxide aerogel	Rhodamine B	I=UV light; PC = 5; CT= 5; AD= 1	Up to 96	[90]
	Orage II	I= UV light; PC = 10; CT= 2; AD= 1	Up to 93	
Nanoparticle gold-iron oxide (maghemite) aerogels	Blue 79 azo dye	I= UV light; PC= 4.85; CT= 3; AD= 0.02 T= 300	Higher than 90	[91]

Irradiation= I, Pollutant concentration (PC, mg L^{-1}),
Contact time (CT, h), Temperature (T, K), Aerogel dose (AD, g L^{-1}).

Conclusion and Future Perspectives

Water pollution and the deterioration of water resources are among the main challenges for our world. Therefore, it is imperative to manage these resources responsibly to avoid and alleviate the escalating effects on agriculture and food supply, health, and the whole environment. The development of materials and water treatment technologies for pollutant removal can help this purpose. Several materials have been devised for capturing and degrading water pollutants by taking advantage of their adsorptive and photocatalytic properties. Aerogels are among this group of materials, which have been studied for environmental applications during the last few decades. These materials can have a large specific surface, a porous structure that can be conveniently tailored, a very active surface, and synthesis processes that can be carried out using environmentally friendly precursors and methods.

Although the formulation, processing, properties and applications of aerogels have been explored by many groups around the world, there are several areas in which there are opportunities for research and development. Fruitful research could focus on synthesis and processing alternatives that can help to overcome some of the drawbacks that have delayed a further development of aerogels. These drawbacks include the need for complex processes (during drying, for instance) and toxic crosslinkers, a sometimes excessively weak structure, with low porosity and surface area, as well as the relatively high costs, which hinder their application. Exploring alternative precursors and more economical and environmentally friendly methods to obtain them, optimizing the synthesis processes, and testing beyond the laboratory scale are some of the research topics that could be pursued. The use of precursors derived from abundant and widely available residues and natural biomass products, which have low or no cost, is one of the most promising options for obtaining aerogels that could offer economical alternatives for large scale water treatment. Processing by conventional and emerging methods should be researched aiming to understand the ways in which the structure and properties of the aerogels obtained can be tailored. The formulation of composites, particularly those in which the reinforcing material could improve not only the removal efficiency but also the mechanical strength and the chemical stability of the aerogels is one of the areas in which there is a need for continuous progress. Increasing the mechanical properties and the stability in acid or alkaline media could also help improving the recycling processes of aerogels as well as their shelf life. Finally, the possibility of developing aerogels that can be regenerated for repeated and efficient use, and that can be disposed in an environmentally safe way, is a topic of interest. In this regard, choosing adequate precursors, solvents and linkers safer for the environment could lead to greener methods.

Most investigations using aerogels are conducted on a laboratory scale and with synthetic wastewater. Therefore, it is essential to look for alternatives to work on a real scale. This implies reducing costs, increasing efficiencies, using materials targeted to remove combinations of pollutants, and exploiting the synergies of different materials, among other pathways. By developing easier methods of synthesis, aerogels can be produced on a large scale, allowing their use in full-scale wastewater treatment or purification plants. On the other hand, synthesizing more versatile aerogels would open the possibility of eliminating a larger portion of the wide variety of pollutants we can find in water. New studies about adsorption with aerogels are required in these cases. Additionally, studies performed combining different aerogels with primary, secondary, and advanced water treatment technologies on a real scale are necessary. There is an opportunity of increasing the efficiency and the useful life of these technologies by exploiting the properties that aerogels possess. Currently there are almost no investigations of this type. Although there are still many unknowns and challenges in using aerogels for pollutant removal, these materials portend a sustainable future for wastewater treatment.

Acknowledgments

The authors would like to acknowledge the support of the Escuela Politécnica Nacional through the group project PIGR-19-05.

References

[1] Mateo-sagasta, J., L. Raschid-sally, and A. Thebo, "Global Wastewater and Sludge Production, Treatment and Use," pp. 15–25, 2015.

[2] Emenike, C. P., et al., "Accessing safe drinking water in sub-Saharan Africa: Issues and challenges in South–West Nigeria," *Sustain. Cities Soc.,* vol. 30, pp. 263–272, 2017.

[3] Peña-guzmán, C., et al., "Emerging pollutants in the urban water cycle in Latin America: A review of the current literature," *J. Environ. Manage.,* vol. 237, no. February, pp. 408–423, 2019.

[4] Rasheed, T., A. A. Hassan, M. Bilal, T. Hussain, and K. Rizwan, "Metal-organic frameworks based adsorbents: A review from removal perspective of various environmental contaminants from wastewater," *Chemosphere,* vol. 259, p. 127369, 2020.

[5] Massoud, M. A., A. Tarhini, and J. A. Nasr, "Decentralized approaches to wastewater treatment and management: Applicability in developing countries," *J. Environ. Manage.,* vol. 90, no. 1, pp. 652–659, 2009.

[6] Rizzo, L., et al., "Consolidated vs new advanced treatment methods for the removal of contaminants of emerging concern from urban wastewater," *Sci. Total Environ.,* vol. 655, no. October 2018, pp. 986–1008, 2019.

[7] Garfí, M., L. Flores, and I. Ferrer, "Life Cycle Assessment of wastewater treatment systems for small communities: Activated sludge, constructed wetlands and high rate algal ponds," *J. Clean. Prod.,* vol. 161, pp. 211–219, 2017.

[8] Gusain, R., K. Gupta, P. Joshi, and O. P. Khatri, "Adsorptive removal and photocatalytic degradation of organic pollutants using metal oxides and their composites: A comprehensive review," *Adv. Colloid Interface Sci.,* vol. 272, p. 102009, 2019.

[9] L. G. Grassi M., Kaykioglu G., Belgiorno V., "Removal of Emerging Contaminants from Water and Wastewater by Adsorption Process," in *Emerging Compounds Removal from Wastewater,* vol. 7, no. 1, 2012, pp. 45–48.

[10] Mlunguza, N. Y., S. Ncube, P. Nokwethemba Mahlambi, L. Chimuka, and L. M. Madikizela, "Adsorbents and removal strategies of non-steroidal anti-inflammatory drugs from contaminated water bodies," *J. Environ. Chem. Eng.,* vol. 7, no. 3, p. 103142, 2019.

[11] Zhu, S., et al., "Adsorption of emerging contaminant metformin using graphene oxide," *Chemosphere,* vol. 179, pp. 20–28, 2017.

[12] Grassi, M., G. Kaykioglu, V. Belgiorno, and G. Lofrano, SpringerBriefs in Molecular Science - Green Chemistry for Sustainability: *Ultrasound Technology in Green Chemistry*. 2012.

[13] Chen, D., et al., "Photocatalytic degradation of organic pollutants using TiO2-based photocatalysts: A review," *J. Clean. Prod.,* vol. 268, p. 121725, 2020.

[14] MiarAlipour, S., D. Friedmann, J. Scott, and R. Amal, "TiO2/porous adsorbents: Recent advances and novel applications," *J. Hazard. Mater.,* vol. 341, pp. 404–423, 2018.

[15] Maleki, H., "Recent advances in aerogels for environmental remediation applications: A review," *Chem. Eng. J.,* vol. 300, no. April, pp. 98–118, 2016.

[16] Vareda, J. P., A. J. M. Valente, and L. Durães, "Heavy metals in Iberian soils: Removal by current adsorbents/amendments and prospective for aerogels," *Adv. Colloid Interface Sci.,* vol. 237, pp. 28–42, 2016.

[17] Zaman, A., F. Huang, M. Jiang, W. Wei, and Z. Zhou, "Preparation, Properties, and Applications of Natural Cellulosic Aerogels: A Review," *Energy Built Environ.,* vol. 1, no. 1, pp. 60–76, 2020.

[18] Lee, J. H., and S. J. Park, "Recent advances in preparations and applications of carbon aerogels: A review," Carbon N. Y., vol. 163, pp. 1–18, 2020.

[19] Liu, W., et al., "Noble metal aerogels-synthesis, characterization, and application as electrocatalysts," *Acc. Chem. Res.,* vol. 48, no. 2, pp. 154–162, 2015.

[20] Kalliola, S., and B. Doshi, "A review of bio-based materials for oil spill treatment," vol. 135, pp. 262–277, 2018.

[21] Wang, L., H. Xu, J. Gao, J. Yao, and Q. Zhang, "Recent progress in metal-organic frameworks-based hydrogels and aerogels and their applications," *Coord. Chem. Rev.,* vol. 398, p. 213016, 2019.

[22] Almeida, C. M. R., M. E. Ghica, and L. Durães, "An overview on alumina-silica-based aerogels," *Adv. Colloid Interface Sci.,* vol. 282, p. 102189, 2020.

[23] Štandeker, S., A. Veronovski, Z. Novak, and Ž. Knez, "Silica aerogels modified with mercapto functional groups used for Cu(II) and Hg(II) removal from aqueous solutions," *Desalination,* vol. 269, no. 1–3, pp. 223–230, 2011.

[24] Ali, Z., A. Khan, and R. Ahmad, "The use of functionalized aerogels as a low level chromium scavenger," *Microporous Mesoporous Mater.,* vol. 203, no. C, pp. 8–16, 2015.

[25] Faghihian, H., H. Nourmoradi, and M. Shokouhi, "Removal of copper (II) and nickel (II) from aqueous media using silica aerogel modified with amino propyl triethoxysilane as an adsorbent: Equilibrium, kinetic, and isotherms study," *Desalin. Water Treat.,* vol. 52, no. 1–3, pp. 305–313, 2014.

[26] Adebajo, M. O., R. L. Frost, J. T. Kloprogge, O. Carmody, and S. Kokot, "Porous Materials for Oil Spill Cleanup: A Review of Synthesis and Absorbing Properties," *J. Porous Mater.,* vol. 10, no. 3, pp. 159–170, 2003.

[27] Gatica, J. M., and H. Vidal, "Non-cordierite clay-based structured materials for environmental applications," *J. Hazard. Mater.,* vol. 181, no. 1–3, pp. 9–18, 2010.

[28] Gawryla, M. D., "Low Density Materials Through Freeze-Drying: Clay Aerogels and Beyond," no. August, p. 221, 2009.

[29] Gawryla, M. D., O. Van Den Berg, C. Weder, and D. A. Schiraldi, "Clay aerogel/cellulose whisker nanocomposites: A nanoscale wattle and daub," *J. Mater. Chem.,* vol. 19, no. 15, pp. 2118–2124, 2009.
[30] Moreno-Castilla, C., and F. J. Maldonado-Hódar, "Carbon aerogels for catalysis applications: An overview," Carbon N. Y., vol. 43, no. 3, pp. 455–465, 2005.
[31] Wang, Y., L. Zhu, F. Zhu, L. You, X. Shen, and S. Li, "Removal of organic solvents/oils using carbon aerogels derived from waste durian shell," *J. Taiwan Inst. Chem. Eng.,* vol. 78, pp. 351–358, 2017.
[32] Zhu, L., Y. Wang, Y. Wang, L. You, X. Shen, and S. Li, "An environmentally friendly carbon aerogels derived from waste pomelo peels for the removal of organic pollutants/oils," *Microporous Mesoporous Mater.,* vol. 241, pp. 285–292, 2017.
[33] Mi, X., G. Huang, W. Xie, W. Wang, Y. Liu, and J. Gao, "Preparation of graphene oxide aerogel and its adsorption for Cu 2+ ions," Carbon N. Y., vol. 50, no. 13, pp. 4856–4864, 2012.
[34] Yang, X., et al., "Highly effective removal of basic fuchsin from aqueous solutions by anionic polyacrylamide/graphene oxide aerogels," *J. Colloid Interface Sci.,* vol. 453, pp. 107–114, 2015.
[35] Bano, Z., et al., "Water decontamination by 3D graphene based materials: A review," *J. Water Process Eng.,* vol. 36, no. December 2019, p. 101404, 2020.
[36] Long, S., et al., "3D graphene aerogel based photocatalysts: Synthesized, properties, and applications," Colloids Surfaces A Physicochem. *Eng. Asp.,* vol. 594, no. February, p. 124666, 2020.
[37] El-naggar, M. E., "Synthesis, Drying Process and Medical Application of Polysaccharide-Based Aerogels," *Int. J. Biol. Macromol.,* 2019.
[38] Wang, Y., Y. Su, W. Wang, Y. Fang, S. B. Riffat, and F. Jiang, "The advances of polysaccharide-based aerogels: Preparation and potential application," *Carbohydr. Polym.,* vol. 226, no. August, p. 115242, 2019.
[39] Mahfoudhi, N., and S. Boufi, "Nanocellulose as a novel nanostructured adsorbent for environmental remediation: a review," *Cellulose,* vol. 24, no. 3, pp. 1171–1197, 2017.
[40] Rong, L., et al., "Facile fabrication of thiol-modified cellulose sponges for adsorption of Hg2+ from aqueous solutions," *Cellulose,* vol. 25, no. 5, pp. 3025–3035, 2018.
[41] Cheng, H., et al., "Multifaceted applications of cellulosic porous materials in environment, energy, and health," *Prog. Polym. Sci.,* vol. 106, p. 101253, 2020.

[42] Chang, X., D. Chen, and X. Jiap, "Chitosan-based aerogels with high adsorption performance," *J. Phys. Chem. B,* vol. 112, no. 26, pp. 7721–7725, 2008.

[43] Vakili, M., et al., "Application of chitosan and its derivatives as adsorbents for dye removal from water and wastewater: A review," *Carbohydr. Polym.,* vol. 113, pp. 115–130, 2014.

[44] Li, A., et al., "An environment-friendly and multi-functional absorbent from chitosan for organic pollutants and heavy metal ion," *Carbohydr. Polym.,* vol. 148, pp. 272–280, 2016.

[45] Salzano de Luna, M., et al., "Optimization of dye adsorption capacity and mechanical strength of chitosan aerogels through crosslinking strategy and graphene oxide addition," *Carbohydr. Polym.,* vol. 211, pp. 195–203, 2019.

[46] Chen, J., Y. Wang, J. Liu, and X. Xu, "Preparation, characterization, physicochemical property and potential application of porous starch: A review," *Int. J. Biol. Macromol.,* vol. 148, pp. 1169–1181, 2020.

[47] Chang, X., D. Chen, and X. Jiao, "Starch-derived carbon aerogels with high-performance for sorption of cationic dyes," *Polymer* (Guildf)., vol. 51, no. 16, pp. 3801–3807, 2010.

[48] Zhang, J. J., S. S. Fang, J. Y. Mei, G. P. Zheng, X. C. Zheng, and X. X. Guan, "High-efficiency removal of rhodamine B dye in water using g-C3N4 and TiO2 co-hybridized 3D graphene aerogel composites," *Sep. Purif. Technol.,* vol. 194, pp. 96–103, 2018.

[49] Ge, J., Y. Zhang, Y. J. Heo, and S. J. Park, Advanced design and synthesis of composite photocatalysts for the remediation of wastewater: A review, vol. 9, no. 2. 2019.

[50] Parale, V. G., T. Kim, V. D. Phadtare, H. M. Yadav, and H. H. Park, "Enhanced photocatalytic activity of a mesoporous TiO 2 aerogel decorated onto three-dimensional carbon foam," *J. Mol. Liq.,* vol. 277, pp. 424–433, 2019.

[51] Zheng, X., S. Xu, Y. Wang, X. Sun, Y. Gao, and B. Gao, "Enhanced degradation of ciprofloxacin by graphitized mesoporous carbon (GMC)-TiO2nanocomposite: Strong synergy of adsorption-photocatalysis and antibiotics degradation mechanism," *J. Colloid Interface Sci.,* vol. 527, pp. 202–213, 2018.

[52] Hasanpour, M., and M. Hatami, "Photocatalytic performance of aerogels for organic dyes removal from wastewaters: Review study," *J. Mol. Liq.,* vol. 309, p. 113094, 2020.

[53] Liu, H., and H. Qiu, "Recent advances of 3D graphene-based adsorbents for sample preparation of water pollutants: A review," *Chem. Eng. J.,* vol. 393, no. January, p. 124691, 2020.

[54] Rego, R. M., G. Kuriya, M. D. Kurkuri, and M. Kigga, "MOF based engineered materials in water remediation: Recent trends," *J. Hazard. Mater.,* p. 123605, 2020.

[55] Du, R., X. Fan, X. Jin, R. Hübner, Y. Hu, and A. Eychmüller, "Emerging Noble Metal Aerogels: State of the Art and a Look Forward," *Matter,* vol. 1, no. 1, pp. 39–56, 2019.

[56] Maleki, H., and N. Hüsing, "Current status, opportunities and challenges in catalytic and photocatalytic applications of aerogels: Environmental protection aspects," *Appl. Catal. B Environ.,* vol. 221, pp. 530–555, 2018.

[57] Pap, Z., et al., "Behavior of gold nanoparticles in a titania aerogel matrix: Photocatalytic activity assessment and structure investigations," Cuihua Xuebao/*Chinese J. Catal.,* vol. 34, no. 4, pp. 734–740, 2013.

[58] Xiang, C., et al., "Synthesis of carboxymethyl cellulose-reduced graphene oxide aerogel for efficient removal of organic liquids and dyes," *J. Mater. Sci.,* vol. 54, no. 2, pp. 1872–1883, 2019.

[59] Zhang, F., Y. H. Li, J. Y. Li, Z. R. Tang, and Y. J. Xu, "3D graphene-based gel photocatalysts for environmental pollutants degradation," *Environ. Pollut.,* vol. 253, pp. 365–376, 2019.

[60] Liu, Q., Y. Liu, Z. Zhang, X. Wang, and J. Shen, "Adsorption of cationic dyes from aqueous solution using hydrophilic silica aerogel via ambient pressure drying," *Chinese J. Chem. Eng.,* 2020.

[61] Medina, M., J. Tapia, S. Pacheco, M. Espinosa, and R. Rodriguez, "Adsorption of lead ions in aqueous solution using silica-alumina nanoparticles," *J. Non. Cryst. Solids,* vol. 356, no. 6–8, pp. 383–387, 2010.

[62] Z. , M. Q. Z. X. M. R. and C. B. Sui, "Green synthesis of carbon nanotube–graphene hybrid aerogels and their use as versatile agents for water purification," *J. Mater. Chem.,* no. Vc, pp. 8767–8771, 2012.

[63] Zhang, W., et al., "Robust shape-retaining nanocellulose-based aerogels decorated with silver nanoparticles for fast continuous catalytic discoloration of organic dyes," *Sep. Purif. Technol.,* vol. 242, no. January, p. 116523, 2020.

[64] Zhang, Y., et al., "Construction of aerogels based on nanocrystalline cellulose and chitosan for high efficient oil/water separation and water disinfection," *Carbohydr. Polym.,* vol. 243, no. May, p. 116461, 2020.

[65] Wan, C., Y. Jiao, S. Wei, L. Zhang, Y. Wu, and J. Li, "Functional nanocomposites from sustainable regenerated cellulose aerogels: A review," *Chem. Eng. J.,* vol. 359, pp. 459–475, 2019.

[66] De France, K. J., T. Hoare, and E. D. Cranston, "Review of Hydrogels and Aerogels Containing Nanocellulose," *Chem. Mater.,* vol. 29, no. 11, pp. 4609–4631, 2017.

[67] Wan, W., R. Zhang, M. Ma, and Y. Zhou, "Monolithic aerogel photocatalysts: A review," *J. Mater. Chem. A,* vol. 6, no. 3, pp. 754–775, 2018.

[68] Awfa, D., M. Ateia, M. Fujii, M. S. Johnson, and C. Yoshimura, "Photodegradation of pharmaceuticals and personal care products in water treatment using carbonaceous-TiO2composites: A critical review of recent literature," *Water Res.,* vol. 142, pp. 26–45, 2018.

[69] Jiang, J., Q. Zhang, X. Zhan, and F. Chen, "A multifunctional gelatin-based aerogel with superior pollutants adsorption, oil/water separation and photocatalytic properties," *Chem. Eng. J.,* vol. 358, no. July 2018, pp. 1539–1551, 2019.

[70] Guan, Y., F. Cheng, and Z. Pan, "Superwetting polymeric three dimensional (3D) porous materials for Oil/Water separation: A review," *Polymers* (Basel)., vol. 11, no. 5, 2019.

[71] B. M. et al., "Polymer-derived ceramic aerogels as sorbent materials for the removal of organic dyes from aqueous solutions," *Int. J. Lab. Hematol.,* vol. 38, no. 1, pp. 42–49, 2017.

[72] Parale, V. G., et al., "SnO2 aerogel deposited onto polymer-derived carbon foam for environmental remediation," *J. Mol. Liq.,* vol. 287, p. 110990, 2019.

[73] Samiey, B., C. H. Cheng, and J. Wu, "Organic-inorganic hybrid polymers as adsorbents for removal of heavy metal ions from solutions: A review," *Materials* (Basel)., vol. 7, no. 2, pp. 673–726, 2014.

[74] Saleem, J., U. Bin Shahid, and G. McKay, "Development of polymeric aerogel using plastic wastes for oil cleanup from wastewater," *Lect. Notes Eng. Comput. Sci.,* vol. 2238, pp. 708–711, 2018.

[75] Álvarez-torrellas, S., et al., "Highly e ffi cient removal of pharmaceuticals from water by well-de fi ned carbide-derived carbons," *Chem. Eng. J.,* vol. 347, no. April, pp. 595–606, 2018.

[76] Guo, D. M., Q. Da An, Z. Y. Xiao, S. R. Zhai, and D. J. Yang, "Efficient removal of Pb(II), Cr(VI) and organic dyes by polydopamine modified chitosan aerogels," *Carbohydr. Polym.,* vol. 202, no. Ii, pp. 306–314, 2018.

[77] Mohseni-Bandpei, A., A. Eslami, H. Kazemian, M. Zarrabi, and T. J. Al-Musawi, "A high density 3-aminopropyltriethoxysilane grafted pumice-derived silica aerogel as an efficient adsorbent for ibuprofen: Characterization and optimization of the adsorption data using response surface methodology," *Environ. Technol. Innov.,* vol. 18, p. 100642, 2020.

[78] Nawaz, M., W. Miran, J. Jang, and D. S. Lee, "One-step hydrothermal synthesis of porous 3D reduced graphene oxide/TiO2 aerogel for carbamazepine photodegradation in aqueous solution," *Appl. Catal. B Environ.,* vol. 203, pp. 85–95, 2017.

[79] Nawaz, M., A. A. Khan, A. Hussain, J. Jang, H. Y. Jung, and D. S. Lee, "Reduced graphene oxide–TiO2/sodium alginate 3-dimensional structure aerogel for enhanced photocatalytic degradation of ibuprofen and sulfamethoxazole," *Chemosphere,* vol. 261, p. 127702, 2020.

[80] Lin, K. Y. A., and W. Der Lee, "Highly efficient removal of Malachite green from water by a magnetic reduced graphene oxide/zeolitic imidazolate framework self-assembled nanocomposite," Appl. Surf. Sci., vol. 361, pp. 114–121, 2016.

[81] Li, Y., R. Zhang, X. Tian, C. Yang, and Z. Zhou, "Facile synthesis of Fe 3 O 4 nanoparticles decorated on 3D graphene aerogels as broad-spectrum sorbents for water treatment," *Appl. Surf. Sci.,* vol. 369, pp. 11–18, 2016.

[82] Chatterjee, A., J. K. Basu, and A. K. Jana, "Alumina-silica nano-sorbent from plant fly ash and scrap aluminium foil in removing nickel through adsorption," *Powder Technol.,* vol. 354, pp. 792–803, 2019.

[83] Chaisuwan, T., T. Komalwanich, S. Luangsukrerk, and S. Wongkasemjit, "Removal of heavy metals from model wastewater by using polybenzoxazine aerogel," *Desalination,* vol. 256, no. 1–3, pp. 108–114, 2010.

[84] Zhao, J., X. Zhang, X. He, M. Xiao, W. Zhang, and C. Lu, "A super biosorbent from dendrimer poly(amidoamine)-grafted cellulose nanofibril aerogels for effective removal of Cr(vi)," *J. Mater. Chem. A,* vol. 3, no. 28, pp. 14703–14711, 2015.

[85] Ma, S., M. Zhang, J. Nie, J. Tan, S. Song, and Y. Luo, "Lightweight and porous cellulose-based foams with high loadings of zeolitic imidazolate frameworks-8 for adsorption applications," *Carbohydr. Polym.,* vol. 208, no. December 2018, pp. 328–335, 2019.

[86] Melone, L., L. Altomare, I. Alfieri, A. Lorenzi, L. De Nardo, and C. Punta, "Ceramic aerogels from TEMPO-oxidized cellulose nanofibre templates: Synthesis, characterization, and photocatalytic properties," *J. Photochem. Photobiol. A Chem.,* vol. 261, pp. 53–60, 2013.

[87] Reddy, D. A., J. Choi, S. Lee, R. Ma, and T. K. Kim, "Self-assembled macro porous ZnS-graphene aerogels for photocatalytic degradation of contaminants in water," *RSC Adv.,* vol. 5, no. 24, pp. 18342–18351, 2015.

[88] Cai, R., et al., "3D graphene/ZnO composite with enhanced photocatalytic activity," *Mater. Des.,* vol. 90, pp. 839–844, 2016.

[89] Liu, W., J. Cai, and Z. Li, "Self-assembly of semiconductor nanoparticles/reduced graphene oxide (RGO) composite aerogels for enhanced photocatalytic performance and facile recycling in aqueous photocatalysis," *ACS Sustain. Chem. Eng.,* vol. 3, no. 2, pp. 277–282, 2015.

[90] Ma, Y., J. Wang, S. Xu, S. Feng, and J. Wang, "Ag 2 O/sodium alginate-reduced graphene oxide aerogel beads for efficient visible light driven photocatalysis," *Appl. Surf. Sci.,* vol. 430, pp. 155–164, 2018.

[91] Wang, C. T., "Photocatalytic activity of nanoparticle gold/iron oxide aerogels for azo dye degradation," *J. Non. Cryst. Solids,* vol. 353, no. 11–12, pp. 1126–1133, 2007.

Chapter 2

Graphene Oxide Aerogels for Oil Spills and Organic Solvents Removal

Nelly Ma. Rosas-Laverde[1] and Alina I. Pruna[2,3,*]
[1]Department of Materials, Escuela Politécnica Nacional, Quito, Ecuador
[2]Institute of Materials Technology, Universitat Politecnica de Valencia, Valencia, Spain
[3]Center for Surface Science and Nanotechnology, University Politehnica of Bucharest, Bucharest, Romania

Abstract

The technological advances also come with a drawback in the air and water pollution. Different absorbents have been used commercially in environmental remediation by treating wastewaters from oil spills and organic solvents. Given that aerogels exhibit improved absorption capacity with respect to powder counterparts, they are lighter and show high regeneration efficiency, a large amount of research studies have been dedicated to such materials. In this respect, graphene aerogels attracted high interest due to the many advantages such as large surface area or hydrophobicity. This chapter highlights recent works in the field of water treatment by oil spill clean up and removal of organic solvents.

Keywords: oil spill, absorption capacity, graphene aerogels

Introduction

Water purification is a global challenger given the necessity to keep developing new and novel materials by using basic and applied science [1].

---

* Corresponding Author's Email: apruna@itm.upv.es.

In: Aerogels: Properties and Applications
Editor: Alina Iuliana Pruna
ISBN: 978-1-68507-788-4

Water pollution is a notable worldwide problem due to increase in industrial wastewater due to the fast technological development [2], urbanization [3] and some environmental accidents caused by crude oil, petroleum products, or toxic organic solvents spills [1, 4]. In order to overcome these issues, natural and artificial materials such as activated carbon, zeolites, fly ash, membranes, macroporous nanocomposites, activated alumina, silica gel, polymeric materials and some waste industrial products have been developed and used to remove pollutants from water. Nevertheless, those materials have various drawbacks including low adsorption capacity, high cost, difficulty in scale-up and low reusability, as well as low environmental compatibility [1, 2, 5, 6]. Some researchers had used expanded graphite, wool fibers, synthetic organic polymers as absorbent materials, nevertheless they show the same problems as mentioned previously, as well as limited practice, low recyclability or even toxicity issues [7]. Furtheremore, sorption capacity is known to depend on properties of sorbent material (e.g., density, pore size, porosity and pore type), as well as operating temperature [8]. Therefore, a novel, efficient, durable and reusable adsorbing material is required to remove or separate varying pollutants including oils, organic solvents and dyes from aqueous media [2].

The pollution from crude oil has increased in the last years due to its transportation from oil extraction zones to oil refineries. In this case the contamination is due to crude oil spills into the sea and it generates serious problems, specially in marine ecosystem [8, 9]. In some research studies, oil spills are considered as "any discharge of liquid hydrocarbons into the environment" [10]. Oil spills involve long-term effects which can cause strong social and economic difficulties and directly affect the human health and ecosystems [10], and they may cause energy loss [11]. It is important to note that some pollutants such as lower molecular weight polycyclic hydrocarbons are very toxic and soluble in the aquatic enviroment [11].

To remove or clean up the oil spills, some investigations focused on the use of various materials and methods including oleophilic sorbents, in situ burning, biological treatment, photocatalytic, chemical, and mechanical methods [10–12]. However, some of these approaches require long time to remove the soluble oil from the deep-water columns [11]. In order to avoid secondary effects or affectations in the environment, the sorption method can be considered as it offers a great, effective, low cost and safe removal alternative [8]. Furthermore, absorption techniques have benefits such as simple fabrication, good recyclability and selectivity [12]. For this purpose, it is necessary to fabricate a superhydrophobic (which means the solid material has water contact angle above 150º) and superoleophilic (which means the solid material exhibits an oil contact angle below 5°) materials which exhibit high absorption capacities to oil spill cleanup and organic solvents.

Under these circumstances, it was observed that carbon-based absorbents became the most popular and novel absorption material lately, due to their high surface area, excellent mechanical properties, chemical stability, reusability, high pore volume, low density and environmental friendliness [12]. Specifically, carbon aerogels can be used for this purpose due to its amphiphilic surface which is created thanks to its three-dimensioanl network porous structure [13].

Carbon Nanomaterial For Aerogels

Generally speaking, carbon materials can exist in all dimensions and can be classified according to type of hybridization bonds of carbon atoms, as can be seen in the Table 1 [14, 15]. These materials can be used in different applications thanks to their various excellent properties.

Table 1. Different types of carbon nanomaterials

Type of carbon material	Hybridization bonds	Dimensionality	Reference
Diamond	sp^3	2D	[14, 17]
Graphite	sp^2	3D	[14, 17]
Graphene **Graphene oxide** **Reduce graphene oxide**	sp^2	2D	[14, 17, 20]
Carbon nanotubes (single and multiwall)	sp^2	1D	[14]
Nanohorns	sp^2	1D	[14]
Nanofibers	sp^2	1D	[14]
Graphene quantum dots	sp^2	0D	[14, 20]
Fullerene	$sp^{2.x}$	0D	[14, 22]
Carbon dots	sp^2	0D	[14, 23]
Nanodiamond **Onion-like carbon**	sp^3 sp^2	0D	[14]
Carbyne	sp	---	[14]
Graphyne	sp – sp^2		[17]

From the carbon nanomaterials, Graphene (G) is known as a two-dimensional (2D) material of honeycomb-arranged carbon atoms which show sp^2 bonds [13, 16, 17]. G exhibits excellent thermal conductivity (5.300 W/mK), it is chemically stable, it has excellent mechanical properties and high electronic mobility, and fast mobility of charge carriers (200.000 $cm^2\ V^{-1}\ s^{-1}$) [18, 19]. Such properties can be improved by adding various hetero-atoms (for example: N, B, S, P and so on). Other properties including its wettability can be tailored, as well [18]. It was reported that G exhibits a sorption capacity for the removal of floating oil as high as 131 $g\ g^{-1}$ [11] thanks to its high exposed surface area of about 433 $m^2\ g^{-1}$ [20]. Moreover, G can exhibit different morphologies and forms including foams, sponges and aerogels [18] which can be used in a variety of applications such as sensors, supercapacitors, catalysts, capacitive deionization systems etc [2, 19]. G can be fabricated by different methods including mechanical exfoliation of graphite, crystal epitaxial growth, chemical vapor deposition and reduction of graphene oxide [21].

Graphene oxide (GO) shows the same structure as G but it is also highly hydrophilic due to presence of hydrophilic groups such as epoxy (–O–), hydroxyl (–OH) and carboxyl acid groups (–COOH) [13, 16], beside containing hydrophobic domains [13]. The GO is a versatile material due to high aqueous dispersability and easy modification conferred by its oxygen groups [24]. One of the most easy, low cost and large-scale methods to obtain GO is Hummers' method [21, 24]. However, thermal exfoliation of graphite oxide was proposed as a great alternative to produce G due to the easiness and low-cost involved. It was observed that the G materials obtained by this approach exhibited elevated surface area and fewer layers [11].

Current Research on Graphene Aerogels

The term aerogel has several definitions: for UPAC, aerogels are defined as "gels comprised of a microporoussolid in which the dispersed phase is a gas" [25]. On the other hand, acccording Eychmüller "an aerogel is a solid with meso- and macropores with diameters up to a few hundred nanometers and a porosity of more than 95 % in which the dispersed phase is a gas" [26].

Given these definitions, the GAs are looked as 3D graphene-based architectures [3] with mesopores which exhibits low density, low thermal conductivity and low dielectric constant, but also shows high specific surface area, high porosity and high acoustic impendence, hydrophobicity, as well as high adsorption capacities [2, 13, 16]. Such properties have made GAs an excellent candidate as sorbent material [8], as well as amphiphilic material [13]. Moreover, their 3D structure offers active adsorption sites to interact with the pollutants and thus, provide a better diffusion into structure and improved performance [3]. It was also observed that GAs show a fast ion/electron transport [13].

The research in GAs has had high impact since Wang, J. and Ellsworth, M. fabricated for the first-time GO and G-based aerogels using "freeze-drying of graphite oxide/water dispersion" in 2009 [31]. On the other hand, Worsley et al. synthetized GA with high electrical conductivity in 2010 [32].

There are more than 14.500 documents related to aerogels. Amongst these, there are about 3.700 researches namely, articles (3.469), reviews (155), proceedings (73), and book chapters (10) covering the use of carbon aerogels in different areas or application such as sensors, photocatalysis, supercapacitors, absorber material and so on. Specifically, there are only 283 documents related to water treatment, 162 that study oil-water separation, 132 for dye adsorption and 36 for organic solvent adsortpion. The counts were obtained using the keywords "graphene aerogel" with "water treatment" or "oil-water separation" or "dye adsorption" or "organic solvent adsortpion" in the ISI Web of Science portal data accessed on October 13, 2021.

Table 2. Properties of different aerogels

Aerogel	Bulk density (g/cm^3)	Specific surface area (m^2/g)	Pore size (nm)	Porosity (%)	Reference
Alumina aerogels	0.17	428	2.2	> 98	[27, 28]
Aluminosilicate aerogels	0.05 – 0.63	150 – 905	0.5 – 50	---	[28]
Carbon aerogels	0.003	500 – 900 2400	---	> 98	[10, 28]
Cellulose aerogels	0.02 – 0.2	100 – 400	1 – 100	---	[29]
rGO aerogels	~0.014	278 – 294	10 – 200	~84	[1]
Polyimide aerogels	0.19	310	29.9	86.0	[30]
Silica aerogels	0.03 – 0.35	600 – 1000	20	~ 99	[28]
Zirconia aerogels	0.05 – 0.45	200 – 900	---	> 97	[28]

To compare varying aerogels, Table 2 offers a view on different properties of aerogels. Thanks to their properties such as good electrical conductivity, flame retardancy, high chemical, thermal and mechanical stabilities, and inherent hydrophobicity, GAs can be used in energy storage/conversion, filtration, microfluidics, sensors, microwave absorption, catalysts, adsorption [10, 33] for water purification [2, 18], which is mostly prominent and concerned application [3], as well as efficient sorbent in solid phase extraction, magnetic–solid phase extraction and solid phase microextraction techniques [20].

Synthesis of Graphene Aerogels

GAs can be obtained by different techniques such as *in situ* assembly, as well as templating, chemical cross-linking, chemical vapor deposition, bidirectional freeze-drying approach, 3D printing, sol-gel, hydrothermal, chemical reduction etc. [2, 28, 33, 34]. In general, carbon aerogels can be fabricated as depending on precursors or methods but it is possible to follow subsequent steps to obtain GAs: synthesis and aging of gels, freeze-drying of gels, and pyrolysis/carbonization [10].

GAs can be obtained by sol-gel method using the resorcinol and formaldehyde as precursors. In this case, the fabrication steps consider the addition reaction in order to include hydroxymethyl groups, then condensation to form methylene and methylene-ether bonds and finally, the cross-linking to form the gel [10]. Other precursor that can be used to form gels by this method are carbon nanotubes and graphene, where the cross-linking is through van der Waals interactions, π–π interactions and/or H- bonds [10].

On the other hand, it is possible to use GO as a precursor to obtain GAs, in this case the aerogel shows pores of variable size [13]. Moreover, GO exhibits a high gelation ability to obtain aerogel with excellent properties such as low weight and density, as well as high porosity which allow reducing agglomerations and can doped with other elements such as nanoparticles, ceramics, and polymers [34].

Other approaches to fabricate GAs are self-assembly process using a template [24], others that are based on chemical or hydrothermal reduction, freeze-drying, chemical vapor deposition and 3D printing [2, 9, 20]. For example, one the first methods to obtain GAs used a GO dispersion, by this approach the hydrophilic surface of GO produces a homogenous aqueous dispersion which is further reduced in order to obtain rGO; then, by different approaches such as hydrothermal one, the sheets can be assembled in a 3D architecture of GAs. GAs are formed during the gelation process of the dispersion which occurs when the balance of the intramolecular forces, as van der Waals and repulsion forces, are broken in the solution and GO sheets are overlapped [6, 20].

Other common method used to fabricate GAs is the chemical vapor deposition. This direct approach uses a metal foam as a template and a carbon source (for example CH_4 gas) to obtain GAs. This approach results in the fabrication of a high-quality 3D carbon nanoarchitectures. For example, nickel or cooper foams were employed as supports for the precipitation of graphene films onto their surfaces. The conditions of the process were ambient pressure and operating temperature of about 1000°C. Furtheron, the supports were removed using chemical etching and finally a GAs were obtained [20].

In the case of hydrothermal approach, homogeneous GO aqueous dispersion is generally used in order to create a 3D network and later, convert it to rGO [20]. The dispersion is heated at 180°C for 12 h using a Teflon–autoclave and produce hydrogels. Once the hold time has passed, the autoclave is cooled down to room temperature and the GAs is rinsed with water. The next step is placing the gels in liquid nitrogen for directional freezing until it is totally frozen and continue with freeze-drying for 24 h [13, 20]. Nevertheless, the GAs obtained without a reducing agent show poor mechanical properties. In order to overcome this issue, it is possible to add different reagents as hydrazine, sodium borohydride, iodic acid, L-ascorbic acid, amine and hydroquinone which can act as reducing agents and improve morphology and mechanical properties as well as accelered the reaction and adsorption capacity [38]. In addition, during the hydrothermal process is possible to add other materials such as ethylene bis(nitrilodimethylene) tetraphosphonic acid (EDTMPA) into the GO dispersion [13], which act as crosslinkers [20], in order to fabricate a 3D microstructure which exhibits amphiphilic properties; in other words, the obtained GAs show hydrophilic and hydrophobic domains. Finally, introduction of gold nanoparticles can be pursued to give catalytic properties to GAs. The GAs fabricate in these conditions and named as EDGA (3D graphene-based aerogel) can be used as an absorbent for different organic liquids. Moreover, this material exhibits recyclability and recoverability properties. The work concluded that the mass-based adsorption capacity of EDGA for tetrachloromethane was ~6800% (wt/wt), for dichloromethane was ~5800%, palm oil was ~4600%, for pump oil was ~3200% [13].

GAs have been shown to exhibit some drawbacks such as brittleness, due to weak Van der Waal's interactions that occur between graphene sheets, or they change from a hydrophilic (GO material) to hydrophobic (rGO material) absorbent material, which means that they may exhibit low absorption and selective capacities for polar solvents [5] and that they may agglomerate [13]. To overcome these issues some researchers used other materials such as organic, synthetic polymers even other carbon materials including carbon nanotubes. In addition, GO dispersion can be mixed with noble metals (as Pd, Ir, Rh, or Pt) or reducing agents ($NaHSO_3$, Na_2S) to improve the mechanical, electrical, and thermal properties [24]. In the research study by Zhao et al. to overcome these issues, inclusion compounds which contains phosphate groups, such as ethylene bis(nitrilodimethylene) tetraphosphonic acid were proposed to fabricate a GAs with hydrophilic surface [13]. Lignocellulose fiber was also proposed as a source to obtain an efficient and recyclable GAs [5].

Water Treatment with Aerogels

In the last twenty years, there have been several oil-spill accidents from offshore platforms and drilling in the oil industry. Many of these accidents were caused by oil spilling from tankers or lacking store tankers, e.g., that provocked by Iraqi soldiers dumping up oil in the sea, or others caused by blowing out or drilling in an oil rig or by sanking of the rig [9, 35, 36]. One of the most significantly accident occurred in the Gulf of Mexico in 2010. In this catastrophe 4.9 million barrels of oil spilled into the sea, and it is considered the world's largest oil spill accident. Nevertheless, more accidents had occurred in other places such as in Bay of Campeche, in the Sundarbans nature reserve in Bangladesh, beach of Kinmen County of Fujian province [12].

All these events or accidents produced severe damage to the marine life, coastal wetlands, fishing and tourims activities, eand they ven caused the death of many workers [9, 35, 36], as the oil spill in the sea got spread over large areas by waves and wind [12].

Furthermore, another secondary effect that is produced by oil spills is the increase of amount of industrial oily wastewater and the leakage of benzene, toluene, cyclohexane, dichloromethane [9].

Considering all these consequences, it got high importancy to fabricate new, advanced, and versatile materials which can act as adsorbent for oils, water-insoluble organic liquids and thus, to reduce such problems. For this purpose, the material should be hydrophobic and oleophilic material, to exhibit high oil-sorption capacity, and to be a low material cost [9]. The absorption capacity of carbon-based absorbents is defined by the relation [12, 37]: *Absorption capacity* = $(m_a$-$m_d)/m_a$, where m_a is the final weight of the material after full absorption and m_d is the initial weight of the material before absorption.

Water Remediation with Non-Templated Graphene Aerogels

Cao et al. have shown a method to functionalize aerogels by hydrothermal process to obtain efficient and recyclable GAs. In this approach, the authors used ethylenediamine (EDA) to reduce GO and to enhance hydrophobicity of aergoel using dodecyltrimethoxysilane (DTMS). The work concluded that obtained GAs show an adsorption capacity of 130–242 g g^{-1} for oils and organic solvents, and of 197–596 mg g^{-1} for dyes. Their study indicated a porous structure of GAs which was formed by pore walls delimited by graphene sheets and a pore size in the range of micrometers. The authors determined that the use of EDA and DTMS allowed the suppression of agglomeration of graphene sheets during the fabrication and enhance the specific area as well as the roughness of GAs [2].

In another study carried out by Hong et al., GAs were functionalized by using rGO aerogel and a solution of trichloro (1H, 1H, 2H, 2H-heptadecafluorodecyl) silane (96%, TCI) in toluene. In this case, the functionalized GAs exhibited a high performance, good recyclability, and mechanical properties, as well as low density and great porosity. Furthermore, this material showed high hydrophobicity which allowed not only to have the high absorption capacity to absorb chlorobenzene, chloroform, pump oil and other organic solvents but also exhibited a good regeneration capability [1].

Bagoole et al. fabricated GAs by hydrothermal approach and funtionalized this absorbent material with trichloro (1H, 1H, 2H, 2H-perfluorooctyl) silane by vapor deposition thus, improving the wettability and hydrophobicity (contact angle more than 150º) properties in order to absorb oil crude and diesel oil. These materials demostranted increased affinity for diesel oil in comparison to the others [21].

One more interesting proposal is by Chen et al. who prepared an absorbent material by using lignin (content in biomass waste) and GAs by hydrothermal approach. In this study they determined that oxygen GO groups are bonding with lignin molecules through intermolecular interactions (for example hydrogen bondings, π-π conjugation and Van der Waal's forces). The results showed that these interactions enabled the the absorbent material for high hydrophobicity, good absorption capacity (167 to 350 times for some oils and solvents) and recyclability [5].

Pruna et al. reported on GAs obtained by combining solvothermal and chemical-induced gelation methods. In this research, the graphene hydrogels were fabricated by using an aqueous slurry of GO nanosheets and vitamin C as reducing agent. In this work, the autors considered the chemical and hydrothermal reduction methods, as well as temperature, duration, and GO oxidation degree to determine their effect on the absorption capacities for oil and organic solvent. The GAs were obtained by the solvothermal method in a temperature range from 65ºC to 165ºC. Then, the hydrogels were frozen at -20ºC. After that, freeze-drying methods at -80ºC under a high vacuum at 0.05 mbar followed by sublimation at 20ºC for 48 h at 0.015 mbar allowed to obtain GAs. The work concluded that the expanded GO aerogels exhibited high absorption capacities [4].

Similarly, Qiu et al. used vitamin C as reducing agent and obtained a GAs by hydrothermal approach. In this research they determined the synergy effect between vitamin C and pH variation, observing that vitamin C allows to obtain GAs with excellent mechanical properties, while by a variation of pH it is possible to control the hydrophobicity of aerogels. In this case, they indicated that GAs obtained with pH 5 had a high absorption capacity for oil from water while that other GAs obtained at pH 11 had a better absorption capacity for methylene blue dye [37].

In the study carried out by Li et al. a solvothermal method was employed to obtain multifunctional magnetite-reduced graphene oxide aerogels (M-RGOs). M-RGOs were prepared by mixing 30 mg of GO with 30 mg of ethylene glycol or 30 mg of triethylene glycol, which acted as reduction agents and solvents to reduce GO. Then 0.2 g of iron acetylacetonate was added to this solution. The final solution was ultrasonicated and stirred, then the mixture was sealed in an autoclave for 24 h at 210ºC and later cooled to room temperature. Finally, M-RGOs were obtained by freeze drying. These absorbent materials showed excellent absorption capacity to separate oil, gasoline, arsenic, and dyes from aqueous media [6].

In order to better understand the influence of pH variation, it is necessary to explain that several external stimuli have been used to improve adsorption/desorption process. Some of these processs could be UV irradiation, electricity, heating, and pH. The latter is the responsible for switching adsorption and desorption process due to the changes produced in the hydrophobicity of surface of absorbent material [7].

On the other hand, the use of copolymers could modify the surface and generate smart materials. For example, Zhu et al. used block copolymer formed by poly(2-vinylpyridine) (P2VP) and polyhexadecyl acrylate (PHA) to modify the pH and create a hydrophobic/oleophilic surface of graphene form (GF). Their results on this smart absorbent material showed a high and fast adsorption capacity for oil and organic solvents, reaching about 196 times of its weight, as well as good recyclability [7].

Figure 1 depicts examples of GAs obtained by hydrothermal technique. The G sheets are clearly observed with a certain stacking as depending on the applied synthetic parameters, as seen in Figure 1a.

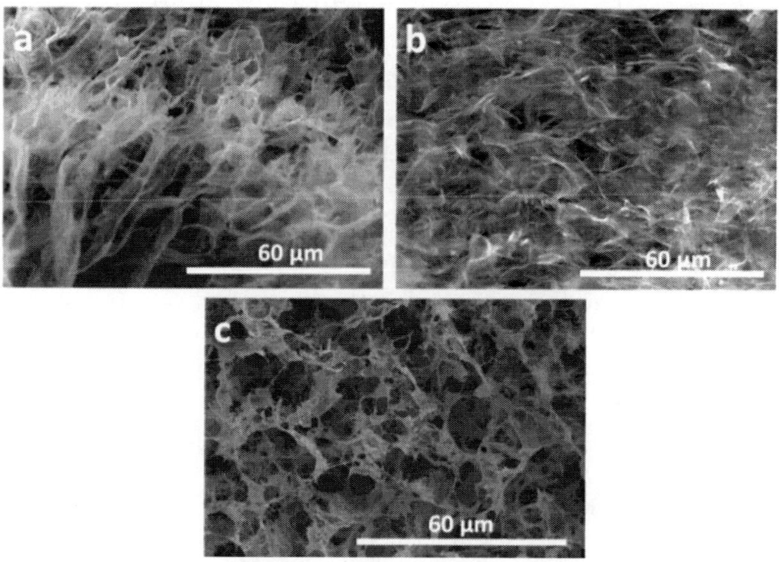

Figure 1. Scanning electron microscopy image of GA: a) obtained by hydrothermal technique and using commercially available GO source nanomaterial; b) obtained by freeze-casting in liquid nitrogen; c) by functionalization with amine groups.

The porosity also depends on these parameters as well as on the choice of freezing technique, as it is observed in Figure 1b. Moreover, the GO properties were indicated to affect greatly the morphology of the obtained GA as well as the addition of varying molecules to functionalize the G sheets, such as amine groups, as depicted in Figure 1c.

Water Remediation with Templated Graphene Aerogels

In order to enhance the mechanical stability, wettability, hydrophobicity or absorption capacities of GAs, some researchers have been fabricating aerogels using other materials such as polystyrene spheres (PS) [24], polyurethane (PU) [39], melamine formaldehyde (MF) microspheres [40], polyvinyl alcohol (PVA) [41], cellulose acetate nanofiber [42] and so on.

For example, Chi et al. obtained tunable hierarchically porous architecture of GAs by self-assembly process using PS (polystyrene spheres) particles which resulted in 3D hierarchical porous graphene aerogels (HGAs). They prepared the aerogel from 2 mg mL^{-1} GO dispersion with 1.44 wt% of PS particle solution, and 30 mg ascorbic acid as reducing agent. In this case, G/GO nanosheets involve the PS particles to form a network by electrostatic repulsions between the PS particles and GO nanosheets. Then, the solution was heat treated and the PS particles were removed with tetrahydrofuran. PSs act as sacrificial templates and determine the porosity and pore size distribution of the microstructure. The obtained 3D macrostructure architecture was analyzed with the presence of PS spheres, and without water. The work concluded that GAs prepared and employed to removal toluene from water reached 93% of the initial uptake [24]. This work is important as toluene is one of the water-insoluble organic solvents that is appear in important in oil spills [9].

Other example is that reported by Han et al, on the fabrication of 3D GAs with hierarchical porous structure by using PS microspheres as sacrificial template. In this case, they fabricated a hierarchical porous graphene aerogel (HPGA) and demonstrated that a favorable situation for gas flow exists thanks to random macropore structure. In order to obtain HPGA structure, they first mixed 5 mg mL^{-1} of GO dispersion with PS dispersion at 9:1 v/v, then freeze-dried for 24 h to obtain PS@GO aerogel. Finally, thermal treatment under inert atmosphere was applied in order to reduce GO and eliminate PS microspheres and obtain HPGA macropores. This process produced a change in color of samples from yellow to black, and did not affect the main morphological structure. The samples obtained in this study were employed to adsorb and monitor air pollutants. In this case, freeze casting was demonstrated as an easy and cheap alternative to fabricate 3D macroporus material by using GO dispersion as a precursor. Finally, the work concluded that the use of PS microspheres as a template allows the obtaining of a hierarchical porous structure.

The scanning electron microscopy images showed that GAs exhibited an interconnected 3D macroporous structure with random pore orientation. The pore size of the structure ranged between 10 μm and 100 μm. Furthermore, it was possible to observe that GO sheets were decorated with PS spheres and that the morphology of the GAs didn't change after the thermal treatment. Nervertheless, cavities in the porous macrostructure were observed after the removal of PS templates [3].

Xiao et al. fabricated multifunctional graphene-based aerogels by using cellulose acetate nanofibers, previously obtained by electrospinning. These nanofibers were added to the GO suspension and homogenized at 14000 rpm for 20 min. Then, the suspensions were freeze-dried at −50ºC for 48 h to obtain nanofiber-modified GA which exhibited good mechanical properties, high adsorption capacity for dyes, oil, and organic solvents from water, as well as a good reusability. In addition, they combined nanofiber-modified GA with polydopamine and polyethyleneimine by co-deposition to enhance the hydrophilicity and oleophobicity properties [42].

In the case of Dai et al., the authors applied freeze-drying approach to develop a new aerogel using GO and PVA, which acts as a cross-linking agent of the 3D porous structure. This new absorbent material exhibited a good stability in non-polar solvent, high selectivity for dyes (specially for the positive charges) and excellent oil/water separation [41].

Furthermore, Cheng et al. preparated GAs using GO and MF microspheres. They used MF microspheres as spacer and pore forming agent. In this case, they obtained a new material with excellent hydrophobicity, low density, large porosity. Their GAs were used to separate organic solvent as acetone, cyclohexane and toluene and oils [40].

The GAs exhibit excellent absorption capacities for oil and organic solvents, as depicted in Table 3. These species are popular pollutants in different areas of our life and industry and they include chlorobenzene, chloroform, crude oil, N-hexane, acetone, methanol, ethanol, decane, toluene, motor oil, pump oil, carbon tetrachloride and so on [1, 5].

The GAs may exhibit different values of absorption capacities which depends on different aspects and in great amount on a combination of high porosity and surface hydrophobicity that these materials exhibit [39]. As it is known, a high water contact angle confers the material with hydrophobic or superhydrophobic surface. Furthermore, graphene-based nanostructures are assembled by interaction between graphene sheets and their external environment that, generally, are obtained through reduction methods, which allow to removal of some oxygen groups [43].

Table 3. Efficiency of GAs for water remediation (selection)

Type of aerogel	Adsorbate	Water Contact angle	Absorption capacity (%)	Ref.
Peanut hull/GA	Oils and organic solvents	141°	3820 – 7933	[44]
Lignin-modified GA	Oils and organic solvents	127°	1800 – 2800	[5]
Melamine formaldehyde microspheres GA	Oils and organic solvents	141°	220 – 280	[40]
Smart 3D porous graphene foam (ss-GF)	Oils and organic solvents	152°	300 – 2000	[7]
Functionalized rGO	Oils and organic solvents	144°	40 – 115	[1]
Expanded graphite (GOx)	Oils and organic solvents	133°	40 – 95	[4]
Nanofiber GA (GNA)	Oils and organic solvents	159°	230 – 760	[42]
Functionalized GA	Oils and organic solvents	144°	130 – 240	[2]
rGO-silica composite	Oil	118 – 131°	7.24 – 11.1	[8]
rGO - candida rugose lipase (rGOB@CRL)	Oil	129.6°	117.67 mg/g	[45]

Other interesting application of GAs is as electrode in capacitive deionization (CDI) technology [19]. For example, Yin et al. fabricated a 3D G/metal oxide hybrid by adding different kinds of nanoparticles such as TiO_2, CeO_2, Fe_2O_3 and Mn_3O_4 in order to improve its capacitance. In this case, they considered that G sheets can easily self-assemble into 3D networks and offer the active sites where NaCl molecules can interact during the deionization process. For example, TiO_2 modified GAs were synthesized by using 60 mg of $TiCl_3$ and 30 mL of 2 mg mL^{-1} GO dispersion as precursors with 10 μL NH_3. To obtain the hydrogel, the hydrothermal process was carried out at 180 °C for 12 h. Finally, the 3D material was obtained by freeze-drying. The Scanning Electron Microscopy images of the GAs confirmed the 3D open pore structure was formed and it showed a pore size between 5–25 μm with walls delimited by G and TiO_2 nanoparticles. In this case, G acted as active site for the nucleation of TiO_2 nanoparticles and allowed that those particles grow up and connect between them. The work concluded that 3D aerogel material exhibits a specific capacitance of 119.7 F g^{-1} and a electro-adsorption capacity of 25.0 mg g^{-1} which confirm the potential use of G/TiO_2 aerogels in CDI application [46].

Conclusion

Graphene nanomaterials have gained increasing research interest in the last years. On the other hand, the environmental protection requires alternatives to restrict the damages caused by oil spills or organic solvents from the wastewaters. The graphene materials have thus, been exploited for this purpose given their ease in processing.

This chapter aimed at presenting the aerogels obtained from graphene nanomaterials by employing different techniques and that were applied for water remediation by removing organic solvents or as a treatment for oil spills. The aerogels were introduced as absorbent materials for this purpose and various carbon nanomaterials have been included in the discussion. Varying characterisctics of aerogels have been highlighted. The focus of this chapter was on the graphene-based aerogels obtained by non-templated and template methods for water remediation. The absorption efficiency of various graphene aerogels was presented for a straightforeword comparison in this respect.

The research results obtained on this topic indicated that high sorption capacity, high recovery rate of pollutant as well as high reusability aerogels are needed to provide with effective sorbent materials for water remediation. The reports on various approaches on graphene demonstrated the viability of such nanomaterial for this purpose.

Important aspects such as the degree of reduction of graphene oxide employed for the synthesis of graphene aerogels, as well as the size and oxidation degree of graphene oxide, together with the functionalization degree with varying molecules, the freeze-casting approach etc. have to be well balanced in order to deliver a well consolidated 3D monolith with good mechanical properties that can be exploited to recover the pollutant (e.g., oil) and to improve the reusability of the absorbent, and to achieve optimum porosity. However, all the steps involved in their synthesis should be simplified and sustainable materials are advised to be employed as the source for the aerogels, while the scalability level also needs to be increased.

Acknowledgments

N. M. R. L. is grateful for the financial support from Escuela Politécnica Nacional.

References

[1] Hong JY, Sohn EH, Park S, Park HS (2015) Highly-efficient and recyclable oil absorbing performance of functionalized graphene aerogel. *Chem Eng J 269:229–235.* https://doi.org/10.1016/j.cej.2015.01.066.

[2] Cao C, Ren J, Lv YK, Ren RP (2021) A functionalized graphene aerogel for efficient water purification. *Green Mater* 9:21–28. https://doi.org/10.1680/jgrma.20.00003.

[3] Han Q, Yang L, Liang Q, Ding M (2017) Three-dimensional hierarchical porous graphene aerogel for efficient adsorption and preconcentration of chemical warfare agents. *Carbon N Y* 122:556–563. https://doi.org/10.1016/j.carbon.2017.05.031

[4] Pruna A, Cárcel AC, Barjola A, et al. (2019) Tailoring the performance of graphene aerogels for oil/organic solvent separation by 1-step solvothermal approach. *Nanomaterials* 9:1–11. https://doi.org/10.3390/nano9081077.

[5] Chen C, Li F, Zhang Y, et al. (2018) Compressive, ultralight and fire-resistant lignin-modified graphene aerogels as recyclable absorbents for oil and organic solvents. *Chem Eng J* 350:173–180. https://doi.org/10.1016/j.cej.2018.05.189.

[6] Li Y, Zhang R, Tian X, et al. (2016) Facile synthesis of Fe 3 O 4 nanoparticles decorated on 3D graphene aerogels as broad-spectrum sorbents for water treatment. *Appl Surf Sci* 369:11–18. https://doi.org/10.1016/j.apsusc.2016.02.019.

[7] Zhu H, Chen D, Li N, et al. (2015) Graphene foam with switchable oil wettability for oil and organic solvents recovery. *Adv Funct Mater* 25:597–605. https://doi.org/10.1002/adfm.201403864.

[8] Thakkar SV, Pinna A, Carbonaro CM, et al. (2020) Performance of oil sorbents based on reduced graphene oxide-silica composite aerogels. *J Environ Chem Eng* 8:103632. https://doi.org/10.1016/j.jece.2019.103632.

[9] Ge J, Zhao HY, Zhu HW, et al. (2016) Advanced Sorbents for Oil-Spill Cleanup: Recent Advances and Future Perspectives. *Adv Mater* 28:10459–10490. https://doi.org/10.1002/adma.201601812.

[10] Ibarra Torres CE, Serrano Quezada TE, Kharissova OV., et al. (2021) Carbon-based aerogels and xerogels: Synthesis, properties, oil sorption

capacities, and DFT simulations. *J Environ Chem Eng* 9:. https://doi.org/10.1016/j.jece.2020.104886.

[11] Wang N, Zhang Y, Zhu F, et al. (2014) Adsorption of soluble oil from water to graphene. *Environ Sci Pollut Res* 21:6495–6505. https://doi.org/10.1007/s11356-014-2504-9.

[12] Gupta S, Tai NH (2016) Carbon materials as oil sorbents: A review on the synthesis and performance. *J Mater Chem A* 4:1550–1565. https://doi.org/10.1039/c5ta08321d.

[13] Zhao Y, Sun T, Liao W, et al. (2019) Amphiphilic Graphene Aerogel with High Oil and Water Adsorption Capacity and High Contact Area for Interface Reaction. *ACS Appl Mater Interfaces* 11:22794–22800. https://doi.org/10.1021/acsami.9b06506.

[14] Satoru Kaneko, Aono M, Pruna A, et al. (2021) Carbon Related Materials.

[15] Kumar S, Nehra M, Kedia D, et al. (2018) Carbon nanotubes: A potential material for energy conversion and storage. *Prog Energy Combust Sci* 64:219–253. https://doi.org/10.1016/j.pecs.2017.10.005.

[16] Li H, Sun C, Liu H, et al. (2019) Aerogels fabricated with origami graphene part I: Preparation and mechanical behavior. *J Alloys Compd* 783:486–493. https://doi.org/10.1016/j.jallcom.2018.12.337

[17] Zhen Z, Zhu H (2018) Structure and Properties of Graphene. In: *Graphene*. Elsevier, pp 1–12.

[18] Rahmani Z, Rashidi AM, kazemi A, et al. (2018) N-doped reduced graphene oxide aerogel for the selective adsorption of oil pollutants from water: Isotherm and kinetic study. *J Ind Eng Chem* 61:416–426. https://doi.org/10.1016/j.jiec.2017.12.041.

[19] Oladunni J, Zain JH, Hai A, et al. (2018) A comprehensive review on recently developed carbon based nanocomposites for capacitive deionization: From theory to practice. *Sep Purif Technol* 207:291–320. https://doi.org/10.1016/j.seppur.2018.06.046.

[20] Nouri N, Khorram P, Sereshti H (2019) Applications of three-dimensional graphenes for preconcentration, extraction, and sorption of chemical species: a review. *Microchim Acta* 186:. https://doi.org/10.1007/s00604-019-3324-x.

[21] Bagoole O, Rahman MM, Shah S, et al. (2018) Functionalized three-dimensional graphene sponges for highly efficient crude and diesel oil adsorption. Environ Sci Pollut Res 25:23091–23105. https://doi.org/10.1007/s11356-018-2248-z.

[22] Zhang Z, Wei L, Qin X, Li Y (2015) Carbon nanomaterials for photovoltaic process. *Nano Energy* 15:490–522. https://doi.org/10.1016/j.nanoen.2015.04.003.

[23] Paulo S, Palomares E, Martinez-Ferrero E (2016) Graphene and Carbon Quantum Dot-Based Materials in Photovoltaic Devices: From Synthesis to Applications. *Nanomaterials* 6:157. https://doi.org/10.3390/nano6090157.
[24] Chi C, Xu H, Zhang K, et al. (2015) 3D hierarchical porous graphene aerogels for highly improved adsorption and recycled capacity. *Mater Sci Eng B Solid-State Mater Adv Technol* 194:62–67. https://doi.org/10.1016/j.mseb.2014.12.026.
[25] IUPAC: Compendium of chemical terminology - Google Académico. https://scholar.google.com/scholar_lookup?title=Compendium of Chemical Terminology&publication_year=1997&author=IUPAC. Accessed 12 Oct 2021.
[26] Ziegler C, Wolf A, Liu W, et al. (2017) Modern Inorganic Aerogels. Angew Chemie Int Ed 56:13200–13221. https://doi.org/10.1002/ANIE.201611552.
[27] Wu L, Huang Y, Wang Z, et al. (2010) Fabrication of hydrophobic alumina aerogel monoliths by surface modification and ambient pressure drying. *Appl Surf Sci* 256:5973–5977. https://doi.org/10.1016/J.APSUSC.2010.03.104.
[28] Walker RC, Potochniak AE, Hyer AP, Ferri JK (2021) Zirconia aerogels for thermal management: Review of synthesis, processing, and properties information architecture. *Adv Colloid Interface Sci* 295:102464. https://doi.org/10.1016/j.cis.2021.102464.
[29] Innerlohinger J, Weber HK, Kraft G (2006) Aerocellulose: Aerogels and Aerogel-like Materials made from Cellulose. *Macromol Symp* 244:126–135. https://doi.org/10.1002/MASY.200651212.
[30] Pei X, Zhai W, Zheng W (2014) Preparation and Characterization of Highly Cross-Linked Polyimide Aerogels Based on Polyimide Containing Trimethoxysilane Side Groups. *Langmuir* 30:13375–13383. https://doi.org/10.1021/LA5026735.
[31] Wang J, Ellsworth M (2019) Graphene Aerogels. *ECS Trans* 19:241–247. https://doi.org/10.1149/1.3119548.
[32] Worsley MA, Pauzauskie PJ, Olson TY, et al. (2010) Synthesis of Graphene Aerogel with High Electrical Conductivity. *J Am Chem Soc* 14067–14069.
[33] Dai D, Zhou Y, Xiao W, et al. (2021) Multiple functional base-induced highly ordered graphene aerogels. *J Mater Chem C* 9:8849–8854. https://doi.org/10.1039/d1tc01985f.
[34] Sultanov F, Daulbayev C, Bakbolat B, Daulbayev O (2020) Advances of 3D graphene and its composites in the field of microwave absorption. *Adv Colloid Interface Sci* 285:102281. https://doi.org/10.1016/j.cis.2020.102281.
[35] Timeline: 20 years of major oil spills - ABC News. https://www.abc.net.au/news/

2010-05-03/timeline-20-years-of-major-oil-spills/419898. Accessed 15 Oct 2021.

[36] Arekhi M, Terry LG, John GF, Clement TP (2021) Environmental fate of petroleum biomarkers in Deepwater Horizon oil spill residues over the past 10 years. *Sci Total Environ* 791:148056. https://doi.org/10.1016/j.scitotenv.2021.148056.

[37] Qiu L, Wan W, Tong Z, et al. (2018) Controllable and green synthesis of robust graphene aerogels with tunable surface properties for oil and dye adsorption. *New J Chem* 42:1003–1009. https://doi.org/10.1039/c7nj03329j.

[38] Wan W, Lin Y, Prakash A, Zhou Y (2016) Three-dimensional carbon-based architectures for oil remediation: from synthesis and modification to functionalization. *J Mater Chem A* 4:18687–18705. https://doi.org/10.1039/C6TA07211A.

[39] Luo Y, Jiang S, Xiao Q, et al. (2017) Highly reusable and superhydrophobic spongy graphene aerogels for efficient oil/water separation. Sci Rep 7:1–10. https://doi.org/10.1038/s41598-017-07583-0.

[40] Cheng Y, Xu P, Zeng W, et al. (2017) Highly hydrophobic and ultralight graphene aerogel as high efficiency oil absorbent material. J Environ Chem Eng 5:1957–1963. https://doi.org/10.1016/j.jece.2017.04.005.

[41] Dai J, Huang T, Tian S qi, et al. (2016) High structure stability and outstanding adsorption performance of graphene oxide aerogel supported by polyvinyl alcohol for waste water treatment. *Mater Des* 107:187–197. https://doi.org/10.1016/j.matdes.2016.06.039.

[42] Xiao J, Lv W, Song Y, Zheng Q (2018) Graphene/nanofiber aerogels: Performance regulation towards multiple applications in dye adsorption and oil/water separation. *Chem Eng J* 338:202–210. https://doi.org/10.1016/j.cej.2017.12.156.

[43] Tan B, Zhao H, Zhang Y, et al. (2018) Amphiphilic PA-induced three-dimensional graphene macrostructure with enhanced removal of heavy metal ions. *J Colloid Interface Sci* 512:853–861. https://doi.org/10.1016/j.jcis.2017.10.092.

[44] Li N, Yue Q, Gao B, et al. (2019) One-step synthesis of peanut hull/graphene aerogel for highly efficient oil-water separation. *J Clean Prod* 207:764–771. https://doi.org/10.1016/j.jclepro.2018.10.038.

[45] Chen M, Wu H, Li Z, et al. (2020) Preparation of reduced graphene oxide porous beads for lipase immobilization and its application for oil adsorption and glycerolysis reaction *in situ*. *Microporous Mesoporous Mater* 294:109920. https://doi.org/10.1016/j.micromeso.2019.109920.

[46] Yin H, Zhao S, Wan J, et al. (2013) Three-dimensional graphene/metal oxide nanoparticle hybrids for high-performance capacitive deionization of

saline water. *Adv Mater* 25:6270–6276. https://doi.org/10.1002/adma.201302223.

Chapter 3

Aerogels for Photocatalytic Degradation of Air and Water Pollutants: Research Trends

Shikha Bansal, Waseem Ashraf, Honey Mittal and Manika Khanuja*
Centre of Nanoscience and Nanotechnology, Jamia Millia Islamia, New Delhi, India

Abstract

With rapid advancement in material science, a new class of materials called as aerogels are being much explored for various types of high-performance applications due to their peculiar characteristics such as significantly low density, tuneable hydrophobicity, poor thermal conductivity, high recoverability, high porosity and high surface area. The presence of organic/inorganic pollutants in the environment generated due to human activities including industrial, agricultural, and domestic entails the development of advanced methodology for the removal of environment pollutants. Aerogel photocatalysts have a proven potential to remove air as well as water contaminants and thus are the promising candidates for the remediation of environment pollutant issues. From the last few decades, with the development of synthesis methodology, many aerogels are being explored for the photocatalytic degradation and energy storage applications ranging from monoliths to films, single components to hybrids, polymers, oxides, and chalcogenides, microporous to mixed porous aerogels. In this chapter, the development of aerogel based photocatalysts and their role in environmental remediation processes for decontamination of water and air have been discussed in detail. Finally, the potential advances in the

---

* Corresponding Author's Email:manikakhanuja@gmail.com.

In: Aerogels: Properties and Applications
Editor: Alina Iuliana Pruna
ISBN: 978-1-68507-788-4

aerogel based photocatalysts are summarized for the future logical examination and commercialization.

Keywords: photocatalysis, air pollution, water pollution

Introduction

Scientists tried to replace the liquid in the pores of the gel with a gas to form an aerogel, in the early 20th century, which is defined by the IUPAC (International Union of Pure and Applied Chemistry) as "a form of gel made up of a microporous dry material with gas as the dispersed phase." [1]. However, the pores of the gel got destroyed during the simple drying process of evaporation, because the surface tension of the interface between liquid and vapor is very strong. After that, Kistler in 1931 [2], achieved success in synthesizing the first aerogel (SiO_2) by using the supercritical drying technique. With the newest advancements in materials science, the aerogels expressed as matters with special properties are extended, and therefore, the methodology for drying is not the only factor for realizing aerogels. As an example, the aerogel of carbon nanotubes can be synthesized directly through the chemical processing of chemical vapor deposition [3]. Various materials (e.g., xerogel and cryogel) having a porous structure square measure are recognized as aerogels. Thus, an aerogel is essentially an engineered, ultra-light and permeable derivative form of a gel in which the liquid part of the gel is supplanted by a gas. For example, aerogel of TiO_2/graphene composite is found to be so light that even a dog's tail grass can be used to hold this aerogel (Figure 1(a)) [4].

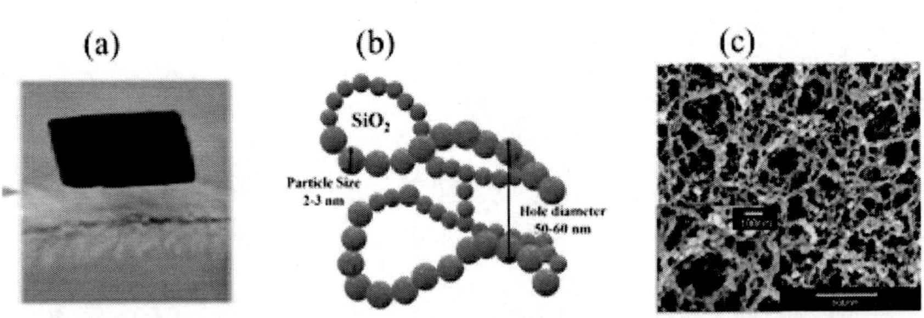

Figure 1. (a) An optical image showing a TiO_2/graphene aerogel sample resting on the dog's tail grass. (reproduced with the permission from [4]) (b) Network architecture of an aerogel. (c) SEM image of the silica aerogel sample [5].

The distinctive properties like ultra-light weight and high thermal resistance of the aerogels are due to their high porousness and intensely little pores. The typical structure of an aerogel material is shown in Figure 1(b) and Figure 1(c) [5].

Aerogels have a fixed shape with fixed volume just like any solid material. However, their density can vary from above 1000 kilogram/meter cube (density of a solid material) to one kilogram/meter cube (air density). Due to such variation in the density of the aerogels, their properties get modified dramatically. The aerogels are also known as frozen smoke, solid air, bluesmoke, or solid smoke because of the light scattering way inside the aerogel and sometimes they also possess translucent nature [6]. Due to the excellent physical and chemical properties, numerous aerogels are fabricated and analysed as potential applicants for diverse commercial applications (Figure. 2), such as thermal insulator, [4], photocatalysts [7], sensor [8], supercapacitor [9], and water purification [10].

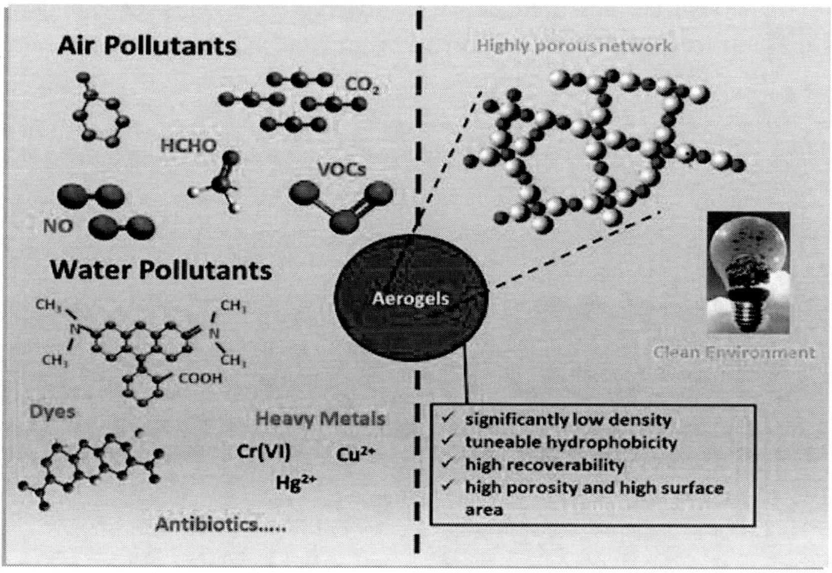

Figure 2. Photocatalytic applications of aerogels including reduction of gaseous pollutants NO_x, CO_2, and photocatalytic degradation of organic pollutants, heavy metal ions and air purification.

The water contamination caused by organic dyes, heavy metals, and air pollutants is one of the most serious environmental problems [11-15]. The water disinfection has attracted extensive interest due to their negative impact on living beings. Photocatalytic degradation using semiconductors is a powerful tool to deal with the issues related to the air pollution and water pollution. The photocatalysis method has the advantages of ease of implementation, and a much lower requirement of energy. However, conventional photocatalytic materials displayed low light adsorption capacity, inefficient recyclability, and high fabrication cost. One of the main drawbacks of photocatalysts in powder is the intricate desorption treatment for recycling use. This necessitates the synthesis of new degradation materials for an efficient and unceasing removal of organic/inorganic toxic waste from the air and water resources. In recent times, aerogels are attracting much of the interest for being the potential candidates for disinfection of water and air pollutants. Aerogels are considered as the ideal photocatalysts due to their high porousness, large surface area, and configurable surface chemistry. In addition, aerogels are capable of floating easily over the aqueous surface due to their low density and tuneable hydrophobicity which facilitates higher light absorption due to maximum exposure, high stability, and recyclability.

Aerogels can be classified into four different categories depending on their appearance, method of synthesis, microstructure and composition. Figure 3 shows the classification of aerogels.

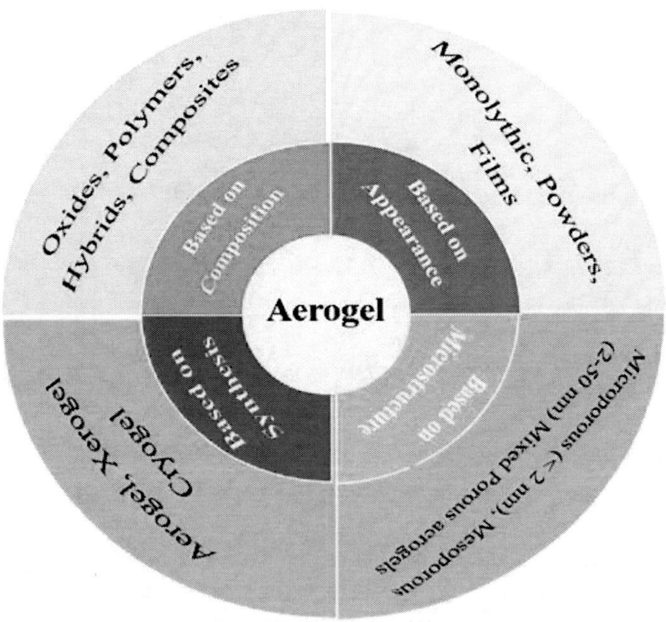

Figure 3. Classification of aerogels on the basis of appearance, preparation method, microstructure and composition.

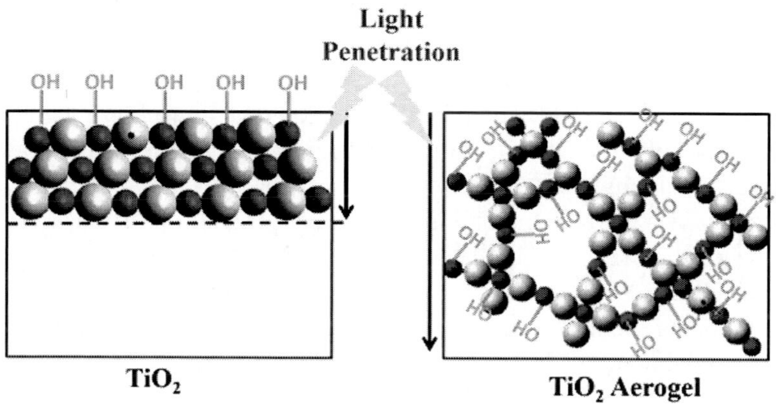

Figure 4. Light accessible surface area in a non-porous and porous material.

An aerogel catalystis characterised as an open cross-linked structure with high porosity (up to 99%), high surface area (up to 2000 mVg), and very low density (down to 0.03 g/cm) which facilitate efficient diffusion, larger light penetration and more active sites (Figure 4). The light accessible area increases by many 1000 times in a porous material as compared to that in a non-porous material. The UV-light accessible surface area of non-porous Degussa (P25) with zero porosity is found to be 1 cm^2 and that for 99% porous TiO_2 aerogel is reported to be 77613 cm^2. This increase in active surface area of an aerogel structure is even more than 1000 times than that of a typical powder. In addition, the recombination of photo-induced electron/hole pairs gets reduced significantly due to the shorter path of migration to the interface between solid and gas [16].

Synthesis of Aerogels

For the synthesis of an efficient photocatalyst, the chosen material should possess emphatically tuneable properties. The physical and chemical properties in addition to the synthesis and processing technology play a pivotal role in the governing the efficiency of aerogels as a photocatalyst. Therefore, the fabrication of an aerogel having a morphology with the availability of a large number of active sites is vital. Aerogels are mostly fabricated from the semiconductor materials having good light-absorbing efficiency in combination with interconnected porous networking of nanostructures which ensures a potentially effective photocatalyst.

Aerogels are synthesized via sol-gel technology (Figure 5). The condensation technique of the aerogel governs the type of structure obtained by an aerogel in addition, the parameters like precursor solution, catalyst used, temperature and the pH value at which the reaction takes place, play a crucial role in structure determination. When the titanium dioxide gel was prepared in an acidic/basic medium, the resultant structure of the material was amorphous, however, a combination of anatase and brookite phase was obtained when the titanium dioxide was prepared in the neutral medium [17]. Aerogel characteristics are significantly influenced by the conditions of gelation including method of drying and gel aging process. Kistler (in 1931) was the first who fabricated the aerogels [2]. He used the supercritical drying technology to remove water from the gel and thus succeeded in preventing the network crashing of the gel and hence the surface area loss. Conventional drying of a gel is carried out in ambient conditions which results in the shrinkage of gel caused by the surface tension at the interface of liquid and vapor. However, if the liquid is held at a pressure above vapor pressure of the liquid and temperature is raised to its critical point then the liquid phase gets converted into gas. The gas so converted can be removed from the gel to get a dry aerogel.

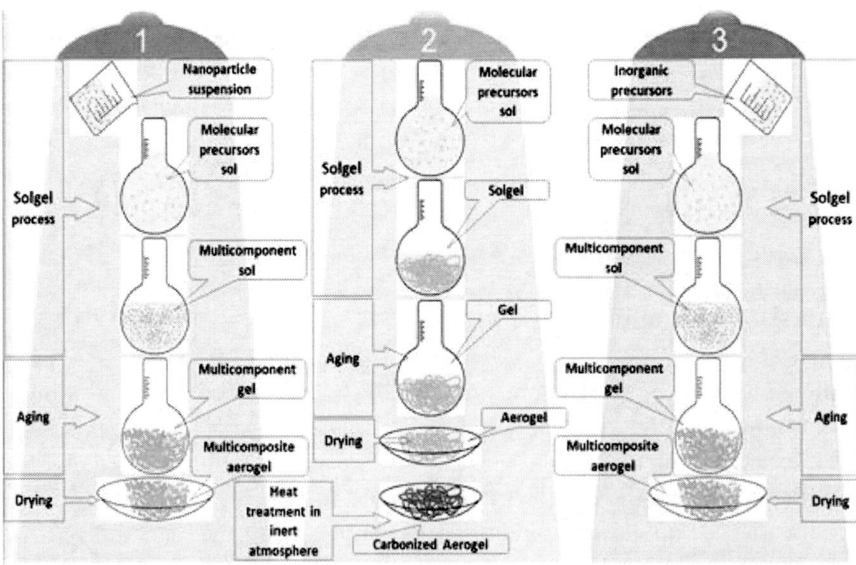

Figure 5. Schematic diagram of general synthesis processes used to prepare aerogel.

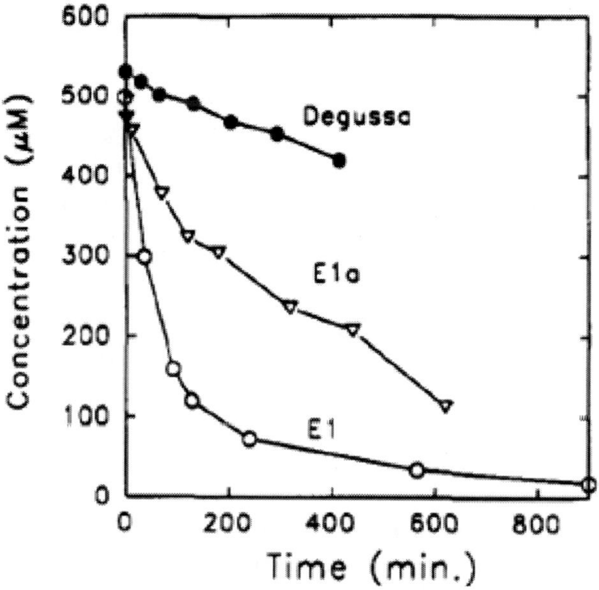

Figure 6. Photocatalytic degradation of salicylic acid over annealed (Ela), nonannealed (El) TiO_2 aerogels and commercial powder (Degussa) (reproduced with the permission from [19]).

Formenti et al., reported the isobutane oxidation over different TiO_2 catalysts. It was observed that the aerogels did not perform well when compared on a mass basis, as nonporous TiO_2 catalysts. [18] However, Dagan et al., [19] by investigating the photocatalytic degradation of salicylic acid, reported an enhanced photocatalytic activity of TiO_2 aerogels than that of Degussa P25 catalyst (Figure 6).

The crystallographic structure of aerogels is a vital factor in governing the photocatalytic response of the catalysts. Moussaoui et al., (2017) [20] compared the photocatalytic performance of TiO_2 aerogels with the commercial TiO_2 over the organic pollutant indigo carmine. TiO_2 aerogel (TA) > commercial Degussa P25-TiO_2 (P25) > TiO_2 xerogel (TX) was the order of catalytic activity for indigo carminede gradation, and higher photocatalytic efficiency of TA was attributed to the higher surface area. However, the poor crystalline quality of TX was responsible for the lower photocatalytic efficiency as compared to P25. Crystalline quality of the photocatalyst facilitates the photocatalytic response and its surface area helps in the catalytic activity. The surface-trapped photo-induced charge carriers lead to the generation of free radicals. However, trapping at the surface and recombination of the charge carriers hinder the formation of oxidant radicals. The amorphous phase of the photocatalyst promotes the trapping and recombination phenomenon [21, 22].

In case of titanium based binary oxide aerogel monoliths, uneven spatial distribution of Ti and formation of cracks is very probable which results in an inactive photocatalyst [23]. Few aerogels are also reported to be photocatalytically inactive even after retaining highly porous structure, interconnected framework and large specific surface area, for example, silica aerogels and binary silica-titania aerogels prepared by supercritical drying by using carbon dioxide [24]. The photocatalytic inactivity of the silica-titania aerogel could be removed through high temperature calcination [25] or by using high temperature and solvents, like, acetone, ethanol etc., during the supercritical drying process or by introducing crystalline nanoparticles.

Pore size and the distribution of pore size also have a significant effect on the photocatalytic activity of the aerogel material. Yao et al., [26] investigated the effect of variation of mesophases of silica-titania aerogels on their photocatalytic response and established that the pores with uniform size are capable of facilitating with larger surface area and hence a greater number of active sites which contributes in the improvement of photocatalytic activity.

Mechanism of Photocatalysis Over Aerogels

Photocatalytic degradation efficiency is highly governed by various characteristics of the photocatalyst like light adsorption efficiency, recombination of photo-inducedcharge carriers, morphology, crystalline natureand hydrophobicity of photocatalyst. Major limiting factors of the efficiency of a conventional photocatalyst (powder form) include low light adsorption capacity due to availability of smaller number of active sites, agglomeration tendency and recyclability. These limitations can be overcome by using a photocatalyst in the aerogel state which is facilitated by an open interconnected frameworks and tunable hydrophobicity. In a typical photocatalytic degradation, the general mechanism involves several steps. First step is the adsorption of pollutants on the surface of the aerogel followed by activation of the aerogel by absorbing light and generation of electron-hole pairs. In the second step, the photo-induced electrons and holes react with oxygen and water molecules present in the atmosphere and leads to the formation of free radicals like superoxide ($O_2^{\bullet-}$) and hydroxyl radicals ($OH^{\bullet}$). The third step involves the decomposition of the pollutant by free radicals into

CO_2, H_2O and other less toxic substances [27]. The photocatalytic degradation mechanism of pollutants over an aerogel is illustrated in Figure 7.

$$Aerogel + h\upsilon \rightarrow Aerogel\,(e^- + h^+)$$
$$Aerogel\,(h^+) + H_2O + OH^\bullet + H^+$$
$$Aerogel\,(e^-) + O_2 \rightarrow aerogel + O_2^{\bullet -}$$
$$O_2^{\bullet -} + H_2O \rightarrow H_2O^\bullet + OH^\bullet$$
$$Pollutant + \frac{\overline{H_2O^\bullet}}{\overline{OH^\bullet}}\!\!\!\!\!\!\underset{O_2^{\bullet -}}{} \rightarrow CO_2 + H_2O + other\ products$$

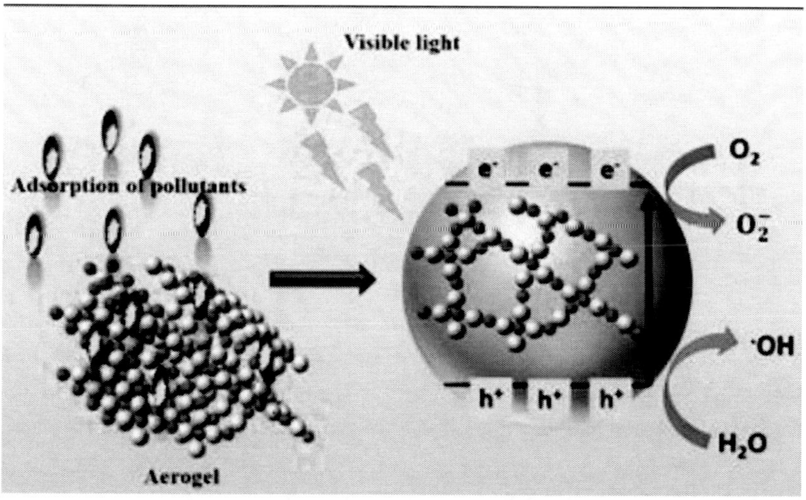

Figure 7. Photodegradation mechanism of pollutants by aerogels.

Photocatalytic Degradation of Water Pollutants

Water is an essential resource for survival of living beings on earth. in conjunction with accessibility, the standard of water is additionally of utmost importance. Over the past few decades, the growing industrial enterprise, domestic waste, use of detergents, fertilizers, pesticides, and drugs, have substantially contributed to environmental deterioration due tothe lack of correct pollution management measures. Broadly, these pollutants can be classified into point and non-point source pollution. Pollution caused from a single place is known as point-source pollution which is identifiable. The non-point source pollution is generated from widely distributed and disconnected places, all at once viz. snowmelt and rain runoff, runoffs from agricultural land, factories etc. and hence it's tough to spot. The various kinds of water pollutants embrace organic pollutants (dyes), inorganic pollutants (heavy metal ions), and pharmaceuticals (antibiotics), as seen in Figure 8.

Figure 8. Sources of water pollution.

Organic Pollutants: Dyes

Dyes are the major pollutants recognized in wastewater owing to significant visibility and affecting the transparency of water by hinderingthe penetration of sunlight. They also restrict the solubility of gas required for water bodies and low biodegradability [28, 29]. In addition, dyes are highly carcinogenic and have adverse effects on humans and aquatic life. These can cause failure to reproductive system, nervous system, liver, and brain, etc. [30, 31] Therefore, there is a need to adopt efficient and economic methodology for decomposition of these organic pollutants.

Several techniques like membrane filtration, adsorption, ion exchange, irradiation, ozonation, electrochemical decontamination, oxidation, and some biological processes like white-rot fungi decolourization, microbial biomass adsorption, and other microbial cultures (Mixed bacterial) [32, 33], are used to elimination the organic dyes dissolved in wastewater. However, these processes involve a considerable consumption of chemicals, energy, disposal of the concentrated sludge along with high processing cost [34]. For this reason, advanced oxidation processes (AOPs) have been proposed to degrade the organic dyes. Among the existing AOPs, heterogeneous photocatalysis which utilizes semiconductor oxides as a photocatalyst has attracted a great interest due to high degradation efficiency and low cost [35-39]. The advantages of heterogeneous photocatalysis include: (i) high rate of degradation, (ii) decomposition of organic compounds to green products, (iii) operatable under room temperature and pressure, and (iv) degradation of complex organic compounds [40]. However, when the most commonly used TiO_2 based photocatalysts are applied for the photocatalytic degradation of the dyes, certain challenges are observed that need to be addressed; viz. insufficient utilization of the solar light due to wide band-gap of TiO_2, fast recombination rate of charge carriers, availability of a smaller number of active sites because of small surface area or low porosity and surface hydrophobicity.

Recent research is therefore focused on development of an efficient photocatalysts having good photocatalytic activity under visible light and slow rate of recombination of charge carriers with larger active sites for disinfection of water [38, 41]. Aerogels are the potential candidates due to their distinctive characteristics such as open pore structure with interconnected framework for mass transport, lightweight, low hydrophobicity and large specific surface area with a greater number of active sites. Due to very low density and controlled hydrophobicity, aerogels tend to float over the surface of water that helps them in more efficient absorption of sun light along with easy recovery even after repeated cycles.

Nowadays many different types of aerogels are being explored for the photocatalytic decomposition of organic dyes from the wastewater. The majority of studies are being carried out on titania aerogels and its hybrids with other organic/inorganic materials for the photocatalytic disinfection of water [42, 43]. Anatase-type TiO_2-C hybrid aerogel was prepared via one-pot sol-gel technique by Shao et al., which showed a photocatalytic degradation efficiency up to 4.23 times more than that of P25 for the degradation of methylene blue (MB) (Figure 9). The higher photocatalytic response of the aerogel was ascribed to the higher absorption of light resulted from porous structure and presence of oxygen vacancies [7].

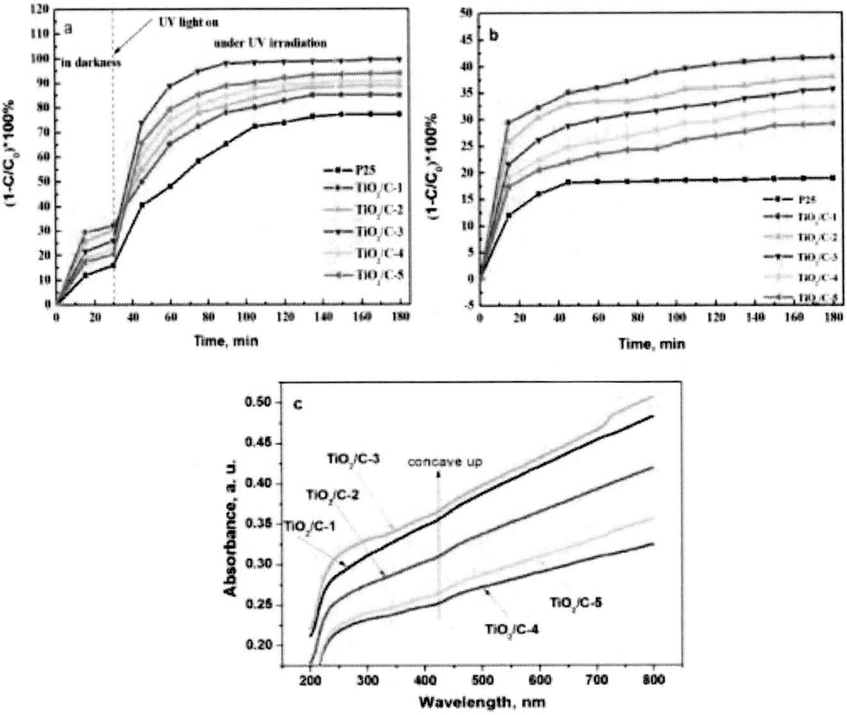

Figure 9. (a) Photocatalytic response of TiO_2-C hybrid aerogel for the degradation of methylene blue (MB) under UV light irradiation; (b) Adsorption of MB in dark and (c) UV-Vis diffuse reflectance spectra of TiO_2-C hybrid aerogel (reproduced with the permission from [7]).

Li et al., fabricated tungsten oxide nanowire-reduced graphene oxide ($W_{18}O_{49}$-RGO) aerogel with low density (6.7 mg/cm^3) and columnar shape. This aerogel was tested for the photocatalytic decomposition of 6 organic dyes: (i) reactive yellow 145; (ii) weak acid yellow G;(iii) reactive black 39; (iv) Rhodamine B (RhB);(v) weak acid black BR; and (vi) methyl orange (MO), under visible light irradiation. All the dyes had shown complete degradation over $W_{18}O_{49}$-RGO aerogel within 25 min. Thus, the aerogel did not show any selectivity of dyes. The recyclability of $W_{18}O_{49}$-RGO aerogel was tested for 30 cycles and was found to maintain a good photocatalysis efficiency with slight variation after 30 cycles for each of the six dyes [44].

Qiu et al., investigated three-dimensional (3D) composite of TiO_2/graphene prepared by using one-step hydrothermal method followed by freeze-drying treatment. Ultra-dispersed TiO_2 single nanocrystals were grown in-situ on the surface of graphene aerogel (GA) (Figure 10). TiO_2 having a mesoporous structure with (001) facets was produced using glucose as the linker and face-controlling agent.

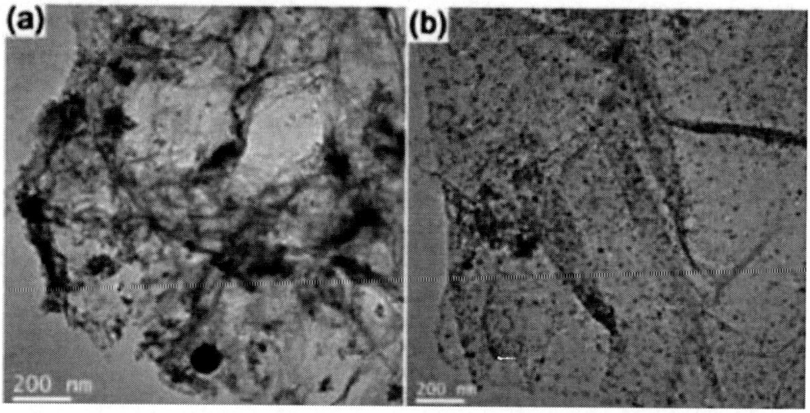

Figure 10. TEM images of TiO2/Graphene aerogel (67 wt %) prepared in the absence (a) or presence (b) of glucose (Reproduced with the permission from [4]).

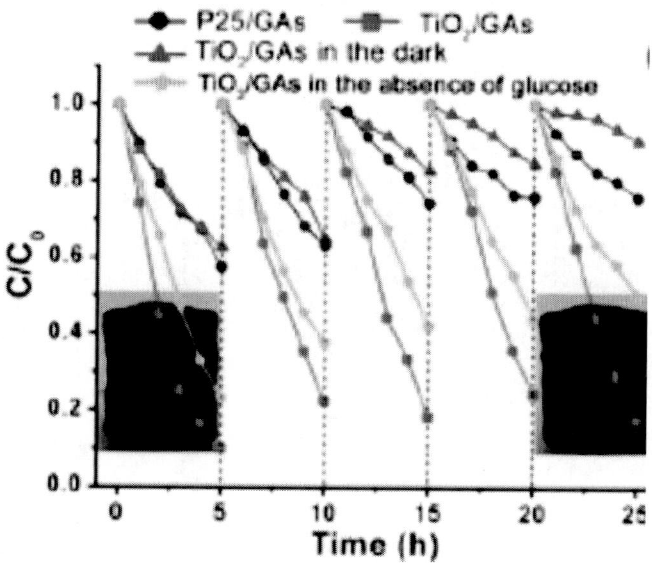

Figure 11. Recyclability test of TiO_2/graphene aerogel and P25/graphene aerogel for the photodegradation of MO under simulated solar light irradiation (reproduced with the permission from [4]).

Table 1. Aerogel photocatalysts used in the photocatalytic degradation of various harmful organic dyes

Aerogel	Synthesis Route	Pollutant	Photocatalytic response	Reference
g-C_3N_4-TiO_2-graphene aerogel	Hydrothermal route followed by freeze drying	RhB	98.4% degradation efficiency in 1h	[46]
CeO_2 nanostructures/ reduced graphene oxide	ultra-sonication assisted chemical method followed by freeze drying	RhB	85% degradation efficiency in 120 min under simulated sun light	[47]
TiO_2–SiO_2	sol– gel process	MB	~ 97% degradation efficiency under UV light	[48]
Nb_2O_5-g-C_3N_4/ reduced graphene oxide aerogel	one-step grinding and calcination followed by chemical reduction self-assembly and freeze drying	RhB	94.8% degradation efficiency under visible light irradiation	[49]
g-C_3N_4/graphene oxide aerogel	hydrothermal co-assembly followed by freeze drying	MO	92% degradation efficiency under visible light irradiation	[50]
Cu_2O/g-C_3N_4/ reduced graphene oxide aerogel	a facile self-assemble method followed by freeze drying	MB MO	96% (MB) and 83% (MO) degradation efficiency under visible light irradiation	[51]
$Cs_{0.33}WO_3/SiO_2$ composite aerogels	Ambient pressure drying	RhB	-	[52]
SiO_2/W_xTiO_2 composite aerogel	hydrothermal deposition method followed by freeze drying	RhB	95% degradation efficiency under visible light irradiation	[53]
SiO_2-TiO_2 aerogel beads	sol-gel method followed by supercritical drying	RhB	95% degradation efficiency under visible light irradiation	[54]
Bi_2WO_6 nanosheets and graphene Aerogel	a facile solvothermal route followed by freeze drying	RhB	99.6% degradation efficiency under visible-light irradiation	[55]
TiO_2-reduced graphene oxide aerogels	hydrothermal method followed by freeze drying	MB	96% degradation efficiency under visible-light irradiation	[56]

Aerogel	Synthesis Route	Pollutant	Photocatalytic response	Reference
g-C_3N_4/graphene oxide aerogel		MO and MB	~90% degradation efficiency under visible-light irradiation	[57]
Nitrogen (N)-doped graphene aerogels- tetraethylenepentamine (TEPA)	hydrothermal method followed by freeze drying	acridine orange (AO)	~ 95% degradation efficiency under visible-light irradiation	[58]
Cu doped Cu_2O/reduced graphene oxide/cellulose hybrid aerogel	in-suit deposition followed by freeze drying	MO	~ 96.8% degradation efficiency under visible-light irradiation	[59]
CuS/cellulose-based aerogel	in-suit deposition followed by freeze drying	MB	~ 96.8% degradation efficiency under visible-light irradiation	[60]
Nitrogen (N) and sulphur (S) codoped graphene aerogels (GA)	hydrothermal method followed by freeze drying	AO	~ 88% degradation efficiency under visible-light irradiation	[61]
BiOI/Silica aerogel	Sono-solvothermal fabrication BiOI over silica-aerogel	MB	~ 96.5% degradation efficiency under visible-light irradiation	[62]
Bi_2MoO_6/reduced graphene oxide aerogel	one-step solvothermal method followed by freeze drying	MB	~ 98.3% degradation efficiency under visible-light irradiation	[63]
Fe_3O_4@TiO_2/SiO aerogel	combined sol-gel and hydrothermal methods	RhB	500 W Xenon-Lamp	[64]
silica aerogel/nano-TiO_2 composite material		RhB	~ 88% degradation efficiency under UV-light irradiation	[65]

3D composites were tested on methylene orange (MO) for the photocatalytic response (Figure 11). High photocatalytic response was attributed to the strong coupling between TiO_2 and GAs, TiO_2 (001) facet characteristics, 3D porous structure, and high electrical conductivity of the TiO_2/graphene composites [4]. In addition, the TiO_2/GAs were found to be very stable and reusable, as confirmed by recyclability test (Figure 11).

Li et al., [45] prepared a cellulose nanofiber/TiO_2 aerogel (CNFT) by using hydrothermal method for the photocatalytic degradation of MB. The composite with composition of 2 ml tetra butyl titanate was reported to have maximum degradation efficiency of approximately 97% under UV irradiation. Table 1 summarises the available reports on the various aerogel photocatalysts that have been used in photocatalytic degradation of many harmful organic dyes.

Removal of Heavy Metal Ions

Heavy metals have long persistence due to non-biodegradability and are highly toxic to the living organisms. Several domestic processes such as use of fossil fuels and sewage sludge, automobile exhaust, incineration of waste, smelting and industrial processes including use of paints, fertilizers and pesticides, ore refining, metal plating, and mining etc. are responsible for the release of heavy toxic metal ions in environment [66, 67]. The presence of heavy metals in water and soil at higher concentrations are detrimental as they can cause serious health hazards like cancer, metal disorder, illness, gastrointestinal problem etc. Recently, heterogeneous photocatalysis has been realized as an economical and reliable technology for the reduction of heavy metal ions. This method utilizes the solar light energy to stimulate the photocatalytic material and produces electron-hole pairs, resulting in disinfection of water and air.

Zheng et al., reported the photoreduction of Cr(VI) over MoS_2/reduced graphene oxide (MoS_2/RGO) aerogel. The photocatalytic efficiency of MoS_2/RGO aerogel in reducing Cr(VI) to harmless Cr(III) was found to be approximately 92% after irradiation for 2 h under visible light [68]. Photocatalytic reduction of Cr(VI) has been realized by using graphene–eosin Y (EY) aerogels under visible light irradiation in 2017 by Yang et al., [69]. These aerogels were prepared by using facile hydrothermal method followed by freeze drying. 3D graphene–EY aerogel was found to possess random open pores and interconnected framework, and visible-light driven photocatalytic activity. The maximum degradation efficiency of about 97% was achieved in 40 min forgraphene–2EY aerogel for the photocatalytic reduction of Cr(VI) under visible light irradiation. A comparative study of the photocatalytic response of graphene–2EY aerogel, graphene–2EY powder and P25-2EY powder for the reduction of Cr(VI) reveals that graphene–2EY aerogel has higher photocatalytic reduction efficiency for Cr(VI) than other two catalysts. Enhanced photocatalytic response established the advantage of 3D aerogel structure for contributing to the photocatalytic activity of composites of graphene and organic compound. In addition, the 3D graphene-organic aerogel shows a good recovery and recyclability which makes this aerogel a potential candidate from practical point of view [69].

ZnS, graphene aerogels (GAs), and ZnS–GAs composites were also investigated by Reddy and co-workers for photocatalytic reduction of Cr(VI) (Figure 12). Under simulated solar light, no significant reduction of Cr(VI) over ZnS and graphene aerogels (GAs) was achieved in 60 min, however, Cr(VI) was found to be reduced by approximately 61% over ZnS-GAs composite. The enhanced photocatalytic response of ZnS–GAs composites was ascribed to the reduced rate of recombination of charge carries and large specific surface area due to the presence of graphene oxide [70].

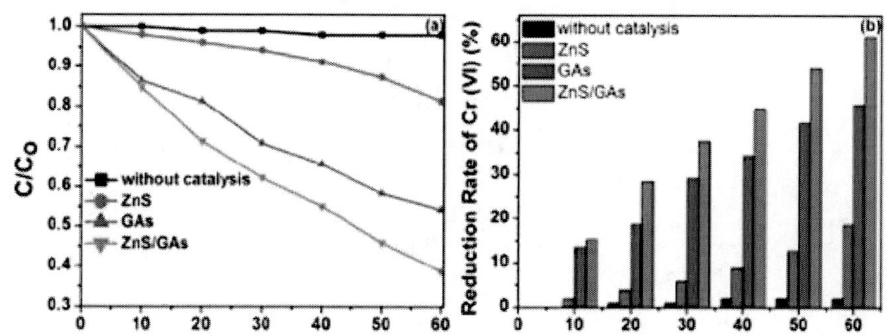

Figure 12. (a) Photocatalytic response of ZnS, graphene aerogels (GAs), and ZnS–GAs composites for the reduction of Cr(VI) under solar irradiation and in the absence of catalysts. (b) The reduction efficiency of ZnS, GAs, and ZnS–GAs composites in the photoreduction of Cr(VI) as a function of time (reproduced with permission from [70]).

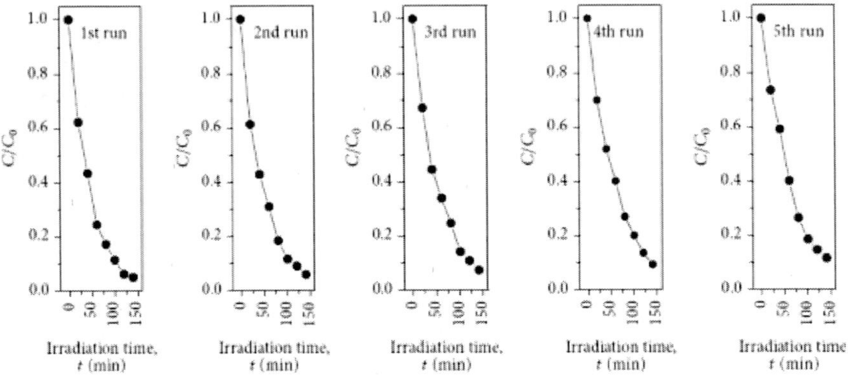

Figure 13. Cyclability tests of $Zn_{0.5}Cd_{0.5}S$/RGO hybrid aerogel for the photocatalytic reduction of Cr(VI) [71].

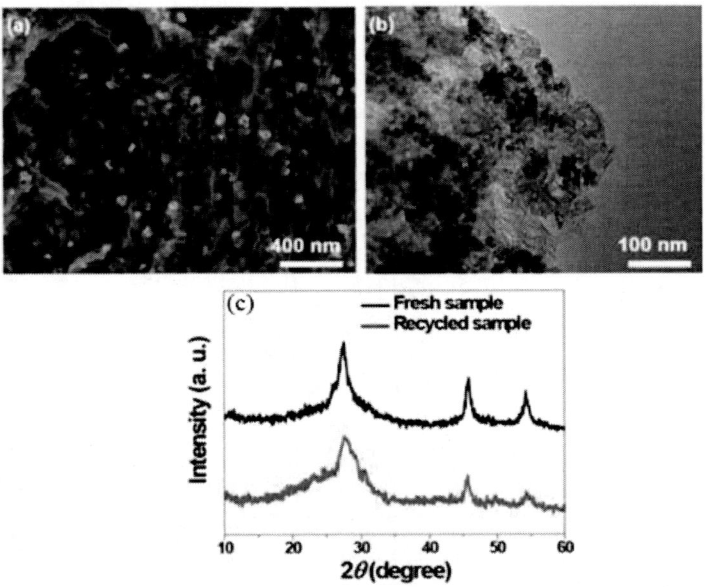

Figure 14. (a) SEM and (b) TEM images of recycled $Zn_{0.5}Cd_{0.5}S$/RGO hybrid aerogel. (c) X-ray diffraction patterns for fresh and recycled $Zn_{0.5}Cd_{0.5}S$/RGO hybrid aerogel. [71].

$Zn_{0.5}Cd_{0.5}S$/reduced graphene oxide ($Zn_{0.5}Cd_{0.5}S$/RGO) hybrid aerogel material was studied by Xiao et al., [71] to investigate its durability and reusability for the photocatalytic reduction of Cr(VI) ions. After the first photocatalytic reaction, $Zn_{0.5}Cd_{0.5}S$/RGO hybrid aerogel was easily separated from the suspension via centrifugation and again dissolved in a new $K_2Cr_2O_7$ solution for the other photoreduction process. It is noticeable that the photocatalytic activity of $Zn_{0.5}Cd_{0.5}S$/RGO hybrid aerogel was maintained till five cycles (Figure 13). The porous structure of aerogel photocatalyst ensured the high recyclability.

The $Zn_{0.5}Cd_{0.5}S$/RGO hybrid aerogel material was found to retain its initial porous structure with uniform distribution of $Zn_{0.5}Cd_{0.5}S$ within RGO even after five cycles as clear from the SEM and TEM images of the recycled $Zn_{0.5}Cd_{0.5}S$/RGO hybrid aerogel (Figure 14(a) and 14(b)). This illustrates the stability and robustness of the aerogel. X-ray diffraction patterns taken for fresh and recycled $Zn_{0.5}Cd_{0.5}S$/reduced graphene oxide ($Zn_{0.5}Cd_{0.5}S$/RGO) hybrid aerogel material confirmed that the hybrid aerogel maintains its porous architecture even after being used for several times for the photocatalytic reduction of Cr(VI) ions (Figure 14(c)) [71].

Pharmaceutical Products (Antibiotics)

Drug items like analgesics, calming drugs, antiseizures, mental medications, hormones, antimicrobials and antilipemic, and so forth are produced for the use for curing human and animal health [72, 73]. Due to their excessive amount present in water resources, these products have become emerging water pollutants that can have irreversible effects on living organisms [74-76]. The main sources of the pharmaceuticals' discharge into the water are pharmacies, hospitals, industries, households, leakages, and leachates from municipal wastewater treatment plants and landfill sites [77]. The conventional methods of wastewater treatment are not efficient in treating such emerging pollutants due to their water solubility and non-biodegradability [78-80].

This alarming situation has led the researchers to search for more efficient and advanced technology i.e., advanced oxidation processes (AOPs) for the removal of such hazardous pollutants for wastewater remediation. The salient feature of AOPs is its basis on the formation of hydroxyl radicals (HO•), which are involved in the disinfection process [81]. Among AOPs, heterogeneous photocatalysis is an emerging transpired technique, which involves redox reactions induced by solar light on the photocatalyst's surface.

The presence of noble metal particles in the semiconductor aerogel matrix significantly enhances its photocatalytic response for the degradation of organic pollutants. TiO_2-Au aerogel composites were investigated by Pap et al., to study the effect of presence and size of gold particles on the photocatalytic efficiency of the material for the degradation of salicylic acid under UV irradiation. The aerogel composite having highest concentration and smallest size of Au-particles showed the best photocatalytic efficiency. The high photocatalytic response of the Au-TiO_2 aerogels was attributed to the increase in crystallized zones and extended crystallization due to smaller Au-particles present in TiO_2 matrix [82].

A similar type of study for photocatalytic degradation of salicylic acid over noble metal doped TiO_2 aerogels was conducted by Sadrieyeh et al., [83]. Highly porous morphological structure of TiO_2 aerogels and strong plasmonic effects of noble metal (Au and Ag) nanoparticles attributes to enhanced photocatalytic activity of Au/Ag-TiO_2 aerogel nanocomposites under simulated solar light. While comparing the photocatalytic response of noble metal doped TiO_2 aerogels, TiO_2 aerogel doped with both Au and Ag nanoparticles showed the best results than Au doped TiO_2 aerogel and Ag-doped TiO_2 aerogel.

Metal-organic aerogels are also being utilized for the degradation of pharmaceuticals [84]. The novel metal organic aerogels named JLUE-MOG-1 and JLUE-MOG-2, containing Fe and Al metal particles, respectively, were prepared by solvothermal method followed by freeze-drying. (Figure 15) for the photocatalytic removal of chlortetracycline hydrochloride (CTC) from water.

An efficient recyclability in addition to the enhanced photocatalytic response was achieved for JLUE-MOG-1 which was attributed to highly porous, easily accessible triazine-moieties and Fe-characteristics (Figure 16).

In another study of photocatalytic degradation of pharmaceutical over metal doped organic aerogels (Me[OA]s) carried out by Bolobajev et al., [85], the aerogel photocatalysts were synthesized by using by-products of Estonian oil shale processing and Fe, Ni, Cu, Zn and Coas dopants. High photocatalytic degradation efficiency was observed for Ni-, Co-, Zn-, Fe[OA] except Cu[OA] and Na[OA]. The poor photocatalytic response of Cu[OA] and Na[OA] was explained on the basis of hypothesis occurrence of both adsorption and photolysis in the absence of sufficient photocatalysis. New energy bands were formed by hybridization with 2p states of O to the delocalized states present in the transition metals. The formation of new energy levels takes place due to these energy bands inside the band-gap which results in the band-gap reduction, facilitating the improved photocatalytic response of the metal organic aerogels.

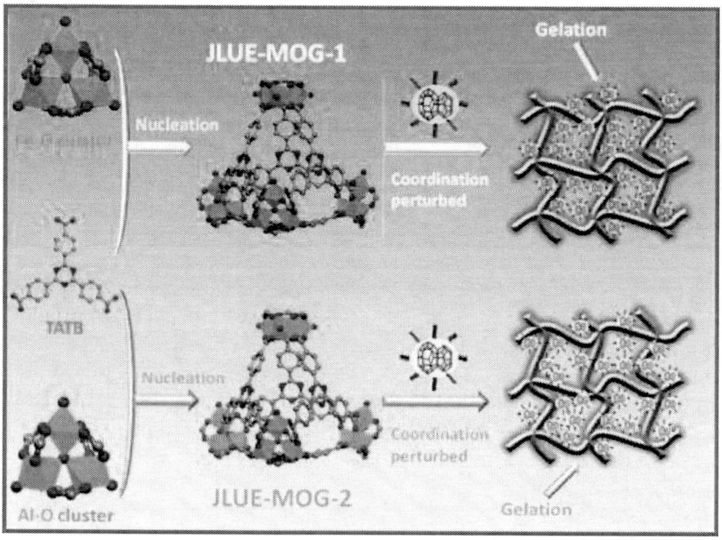

Figure 15. Schematic diagram of the synthesis of metal organic aerogels named JLUE-MOG- 1 and JLUE-MOG-2 aerogels (reproduced with the permission from [84]).

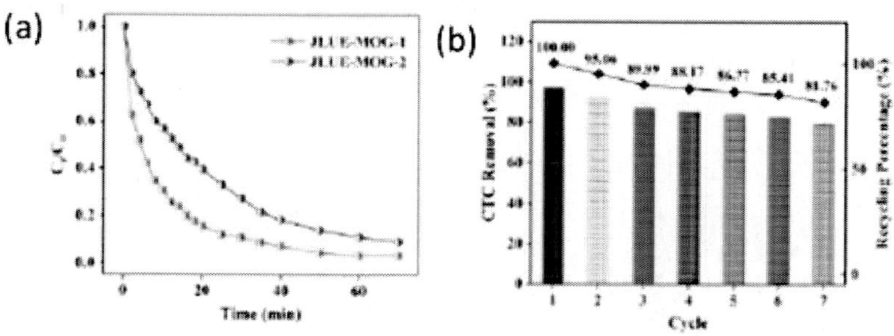

Figure 16. (a) Photocatalytic removal of chlortetracycline hydrochloride (CTC) by metal organic (JLUE-MOG) aerogels (b)Cyclabilitytests of JLUE-MOG-1 aerogels for the photocatalytic removal of CTC. (reproduced with the permission from [84]).

Liu et al., [86] prepared silver phosphate/graphene oxide aerogel microspheres (SGAMs) with radially oriented microchannels using electrostatic spray technique followed by freeze-drying, lyophilization and crosslinking (Figure 17) and tested this aerogel for the removal of bisphenol A (BPA) using photocatalysis.

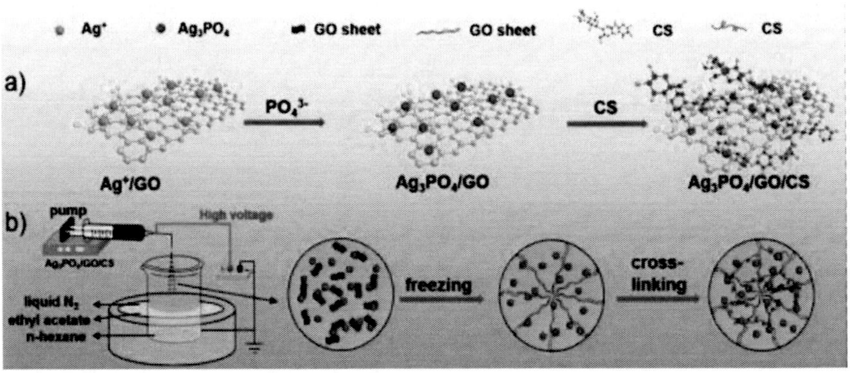

Figure 17. Schematic representation of the synthesis of (a) Ag_3PO_4/GO/CS suspension and (b) porous SGAMs with radially oriented microchannels. (reproduced with the permission from [86]).

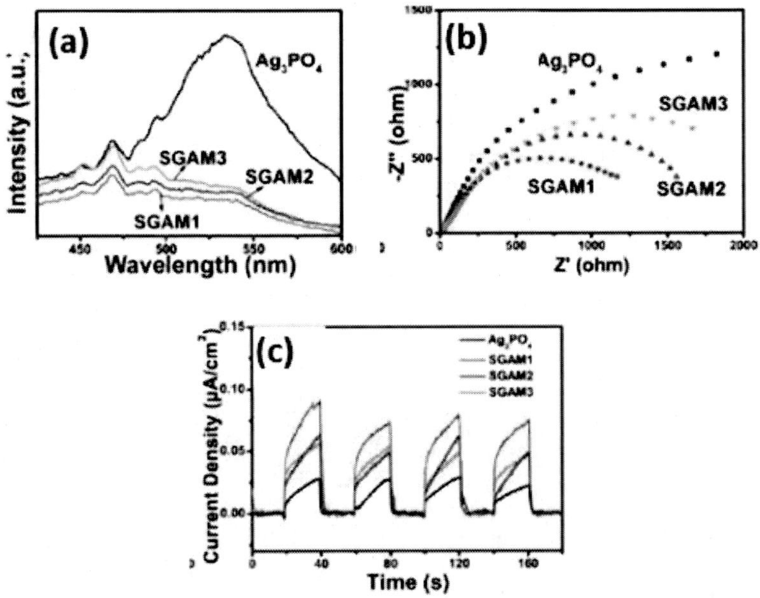

Figure 18. (a) Photoluminescence spectra, (b) EIS curves, and (c) photocurrent transient responses of Ag_3PO_4 and SGAMs (reproduced with permission from [86]).

The radially oriented microchannels of SGAMs and the chemical bonding between GO and Ag_3PO_4 contributed to the acceleration of transfer of charge carriers and hence reduced the rate of recombination of charge carriers, which is confirmed from the weak photoluminescence peak intensities, small semicircle diameters of EIS curves and enhanced current density of Ag_3PO_4 and SGAMs (Figure 18).

In addition, the presence of GO in aerogel composite ensures the stability and recyclability of the catalyst (Figure 19).

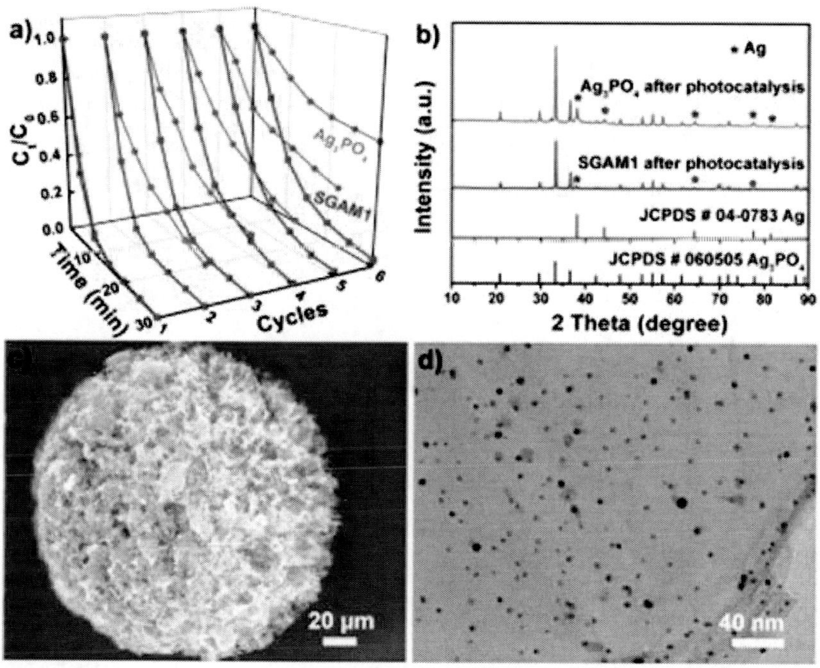

Figure 19. (a) Cyclability tests of Ag_3PO_4 and SGAM1 for the photodegradation of BPA. (b) XRD patterns, (c) SEM image, and (d) TEM image of SGAM1 after 6 degradation cycles(reproduced with permission from [86]).

The different concentrations of BPA were efficiently degraded by SGAMs under visible light. The effect of pH value, temperature and ionic strength on the photocatalytic degradation of BPA by SGAM was also investigated (Figure 20). It was observed that photocatalytic degradation rate deteriorates in highly alkaline and highly acidic solution. However, the increased temperature favours the degradation rate. The high ionic strength also has an adverse effect on photocatalytic degradation rates.

SGAMs are also found efficient in the photocatalytic degradation of the emerging pharmaceutical pollutants: ibuprofen (IBU), sulfamethoxine (SMZ) and atrazine under visible light irradiation (Figure 21). The better photocatalytic performance of SGAMs makes them potential candidates for the removal of water pollutants.

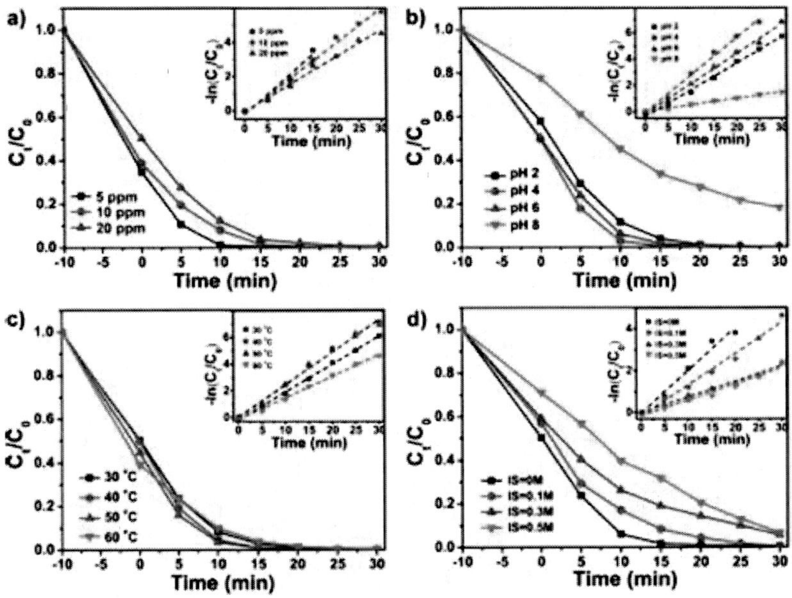

Figure 20. (a) Photodegradation of different initial concentrations of BPA over SGAM1. Effect of (b) pH value, (c) temperature, and (d) ionic strength (IS) on photodegradation of BPA over SGAM1. Insets are their rate kinetic curves(reproduced with permission from [86]).

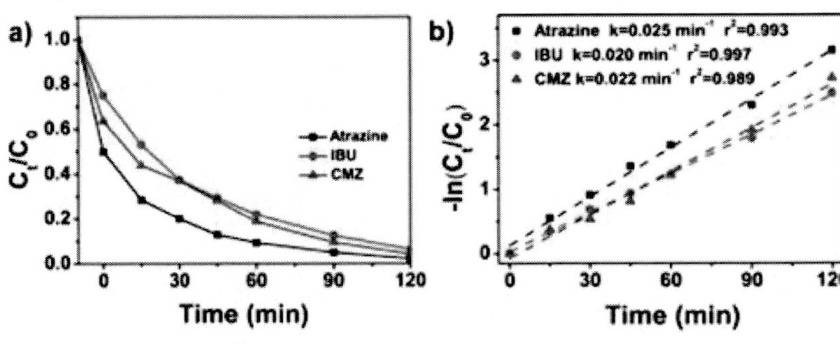

Figure 21.(a) Photocatalytic response of SGAM1 under visible light irradiation for the degradation of various antibiotics: ibuprofen (IBU), sulfamethoxine (SMZ), and atrazine and (b) Corresponding rate kinetics(reproduced with permission from [86]).

Table 2. Aerogel photocatalysts used in the photocatalytic degradation of various drugs

Aerogel	Synthesis Route	Pollutant	Photocatalytic response	Reference
Z-scheme Polyimide (PI)/AgBr@Ag aerogel	Precipitation method followed by supercritical drying	Oxytetracycline	~ 99.8% degradation efficiency under visible light	[87]
3D BiOCl/ RGO aerogel	Hydrothermal method followed by freeze drying	Oxytetracycline	~ 93.3% degradation efficiency under 300 W xenon lamp	[88]
Tungstate doped TiO_2-SiO_2 aerogels	Sol-gel method followed by heat drying	Methamphetamine	Almost complete degradation under simulated sunlight	[89]
TiO_2 Aerogels	sol-gel method followed by supercritical drying	Salicylic acid	~65% degradation efficiency by using 85-W Hg lamp	[14]
g-C_3N_4-agar hybridaerogel	heating-cooling polymerization process	Metronidazole (MNZ), ciprofloxacin hydrochloride (CIP)	~95% (MNZ), ~92% (CIP) under simulated-sunlight	[90]
Rod-shaped Carbon aerogel-assisted CdS nanocomposite	hydrothermal process followed by freeze-drying	Phenol	~95% degradation efficiency irradiated by xenon light of 500W	[91]
3D g-C_3N_4@cellulose aerogel	Cross-linked blended polyester fibres run through the porous cellulose aerogel containing nanosheet g-C_3N_4	Sulfaquinoxaline sodium (SQX),	100% degradation efficiency under sunlight irradiation	[92]
syndiotactic polystyrene (s-PS) and N-doped TiO2 monolithic aerogel	supercritical CO_2 extraction of N-doped TiO_2 /s-PS gels	Phenol	~48% degradation efficiency under irradiated with UV light	[93]
TiO_2-SiO_2 binary aerogels	sol-gel process followed by supercritical drying	Phenol	~50% degradation efficiency irradiated by using ultraviolet lamp (λ=253.7nm, 15W)	[94]

Table 2. (Continued)

Aerogel	Synthesis Route	Pollutant	Photocatalytic response	Reference
TiO_2 aerogels	sol-gel method followed by freeze drying	salicylic acid	~60% degradation efficiency under 85W Hg lamp	[95]
Polyimide (PI) aerogel cross-liked through multi-walled carbon nanotube (MWCNT)	gelation of chemically imidized polyimide in N-methylpyrrolidinone (NMP) solution, followed by ethanol supercritical drying	oxytetracycline	~60% degradation efficiency under UV-light source (250 W)	[96]
Titania–silica composite aerogel	sol–gel method followed by supercritical drying	Salicylic acid	~30% degradation efficiency by using 150 W Hg vapor lamp	[97]
TiO_2 aerogel	sol–gel method followed by supercritical drying	Phenol	~92% degradation efficiency under UV irradiation by using two high-pressure mercury lamps	[98]
Organic aerogels	polycondensation of resorcinol and formaldehyde in water, using Fe, Ni, and Co acetates as catalytic precursors	Amitrole	~71% degradation efficiency under UV light by using a low-pressure mercury vapor lamp	[99]
3D reduced graphene oxide/TiO_2(RGOT) aerogel	hydrothermal method followed by freeze drying	carbamazepine (CBZ)	~99% under UV light	[100]
TiO_2 aerogels	sol-gel method followed by supercritical drying	salicylic acid	~85% degradation efficiency by using 85W Hg lamp	[101]
TiO_2-nanosheets/Pt/GO aerogel	hydrothermal method followed by freeze-drying	Chlortetracycline	~45% degradation efficiency under visible-light	[102]

Various aerogel based photocatalysts have been successfully implied for the photocatalytic degradation of pharmaceuticals. Table 2 presents the list of different aerogel photocatalysts that have been used in the photocatalytic degradation of several hazardous pharmaceuticals.

Photocatalytic Response of Aerogels in Air Cleaning

Air Pollution

Air pollution is one of the major concerns for living beings and so drawing the interest of health research and environmental regulatory communities. Because the indoor, as well as the outdoor air quality, is related to the despondency and lethality [103]. Presently, the major cause of air pollution is combustion of fuel for industrial and domestic production, as depicted in Figure 22. Also, the consumption of fuel for production and transport, pesticides, fertilizers, and aerosol sprays is user dependent [104, 105]. The condition of vehicles and the type of fuel used have a significant effect on automobile discharge. A facund decrease in ambient air pollutants has been recorded by compressed natural gas (CNG) instead of diesel or petrol in vehicles in Delhi, India. According to the recorded data of 1995 to 2001, the average concentrations of suspended particulate matter (PM), carbon monoxide (CO), and NO_x is reduced from 405 Ag/m^3 to 347 Ag/m^3, from 4681 Ag/m^3 to 4197 Ag/m^3 and from 34 Ag/m^3 to 36 Ag/m^3, respectively [106].

Six criteria air pollutants aredeclared by National Ambient Air Quality Standards [107]. 5 primary pollutants are particulate matter, SO_2, NO_2, CO and Pbwhich are directly released into atmosphere. Ozone is a secondary pollutant that comes under the category of pollutants that are formed in the lower atmosphere due to the chemical reactions between sunlight and the other pollutants present in the atmosphere like hydrocarbons and nitrogen oxides. Another category of hazardous air contaminants consists of volatile organic compounds (VOCs) with significant vapor pressures, such as hydrocarbons (aldehydes, organic acids, ketones) as well as hydrocarbons containing chlorine (halogens), sulphur, nitrogen, and fluorine [108].

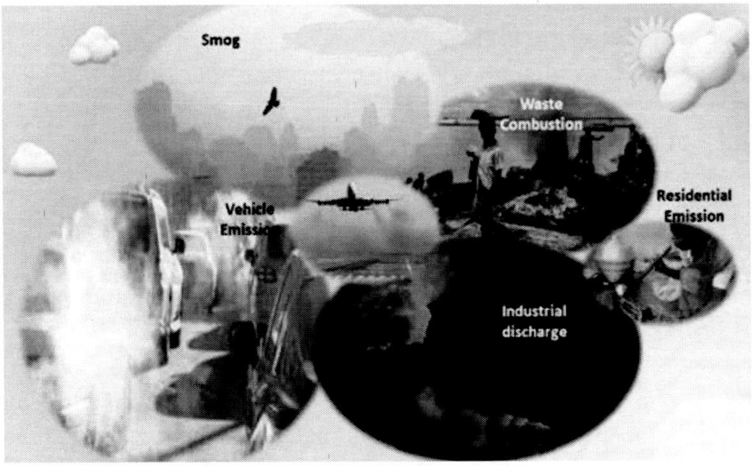

Figure 22. Sources of air pollution.

Six criteria air pollutants are declared by National Ambient Air Quality Standards [107]. 5 primary pollutants are particulate matter, SO_2, NO_2, CO and Pb which are directly released into atmosphere. Ozone is a secondary pollutant that comes under the category of pollutants that are formed in the lower atmosphere due to the chemical reactions between sunlight and the other pollutants present in the atmosphere like hydrocarbons and nitrogen oxides. Another category of hazardous air contaminants consists of volatile organic compounds (VOCs) with significant vapor pressures, such as hydrocarbons (aldehydes, organic acids, ketones) as well as hydrocarbons containing chlorine (halogens), sulphur, nitrogen, and fluorine [108].

Available Remediation Technology for Air Pollution

Many conventional technologies are available ranging from incinerators (thermal or catalytic oxidation) [109] to methods of separation like, filter systems [110, 111], adsorption [112-114] and liquid absorption [115]. However, these conventional techniques involve significantly high operating costs. These separation methods produce a lot of waste streams consisting of contaminants that are separated; additional waste is generated due to cleaning of filters/separation devices, which necessitates the eradication of generated waste. Therefore, it is imperative to invent an economical method with reduced waste generation, for the detoxification of these pollutants that are widely spread in the air.

Throughout the research and development of new technologies, heterogeneous photocatalysis has demonstrated for being a successful alternative or addition to existing degradation processes [116]. Zhang et al., investigated the photocatalytic degradation of automobile exhaust over TiO_2-SiO_2 aerogel under UV-irradiation. TiO_2-SiO_2 aerogel worked well in degrading the various components (CO, CO_2, HC and NO_x) of automobile exhaust. The maximum photocatalytic degradation of NO_x over TiO_2-SiO_2 aerogel was achieved among all components of automobile exhaust under investigation [117].

Nitrogen oxides (NO_x)

Nitrogen oxides are one of the main air pollutants which also serves as a precursor for chemical smog and ozone at ground level. They are a family of highly reactive poisonous gas. Exposures to these gases can cause airway responsiveness, lung injury, reduced immunity, and respiratory infections. Traffic emissions and fuel combustion are two major sources of nitrogen oxide gases. In addition to the regulatory measures taken for controlling air pollution, photocatalysis has been recognized as a promising tool for low-concentration NO_x removal. Various photocatalysts such as metal oxides (Bi, TiO_2, BiOX: X = Cl, Br, or I) [118-124], perovskites ($MTiO_3$: A = Ba, Sr or Pb) [125-127] and metal-free carbon-based compounds like graphitic carbon nitride having high photodegradation response have been investigated for the removal of nitrogen oxides [128, 129]. However, many photocatalysts lag in efficiency due to a smaller number of available active sites or high recombination rates. In recent years, aerogel-based materials are leading to the widening of the application as photocatalyst due to their highly porous cross-linked networking and very low density.

Various studies are carried out on graphitic-carbon nitride (g-C_3N_4) based aerogels to remove NO_x using photocatalysis. Hu et al., [130] investigated g-C_3N_4 modified with perylene imide (PI) and graphene oxide (GO) aerogel as photocatalyst for the removal of nitric oxide (NO) under visible-light. An excellent photocatalytic response of PI-g-C_3N_4-GO aerogel was observed with degradation efficiency of 66% for NO removal which was attributed to strong light absorption capacity of PI-g-C_3N_4 aerogel and reduction in charge recombination rate facilitated by GO. The recyclability of the catalyst was tested for multiple cycles and the photocatalyst was found to be quite stable. The visible light absorption by photocatalyst leads to the generation of electron-hole pairs. The electrons from the conduction band (CB) of PI recombined with holes in the valence band (VB) of g-C_3N_4. Then the holes in the VB of PI oxidized NO to NO_3 and the electrons in the CB of g-C_3N_4 reduced oxygen to $\cdot O^{2-}$ and $\cdot OH$. Intermediate product formed in this process was NO_2.

An aerogel with composition g-C_3N_4 quantum dots (CNQDs) modified with GO and $InVO_4$ also gave the similar photocatalytic response in NO removal (ppb-level) under visible-light irradiation [131]. The highest removal efficiency of 65% was achieved for CNQDs/GO-$InVO_4$ aerogel for NO removal with negligible loss of efficiency even after four cycles.

The enhancement of photocatalytic activity of g-C_3N_4 for NO removal by two step modification was also confirmed by Liu et al., [132]. The photocatalytic NO removal was performed over g-C_3N_4 modified with graphene aerogel and $AgVO_3$. A clear effect of slow recombination rate of electron-hole pair due GO was confirmed in $AgVO_3$-g-C_3N_4-graphene hybrid aerogel.

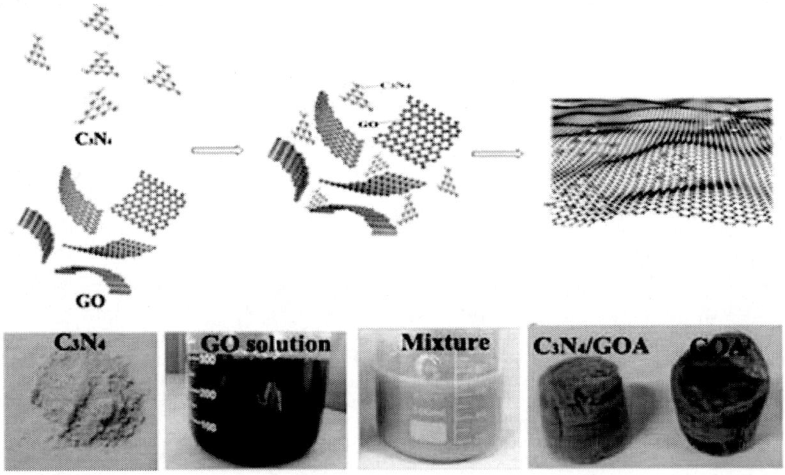

Figure 23. Schematic illustration of synthesis of C_3N_4/graphene oxide aerogel (C_3N_4/GOA) via freeze-drying self-assembly route (reproduced with permission from [133]).

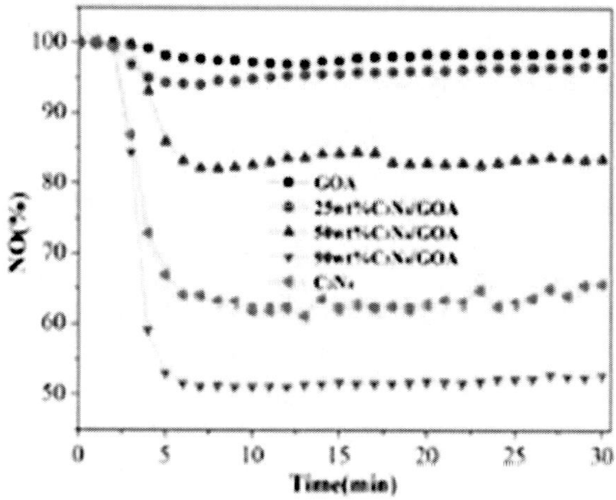

Figure 24. Photocatalytic response of C_3N_4/GO aerogel with different concentration of C_3N_4 for removal of NO (reproduced with permission from [133]).

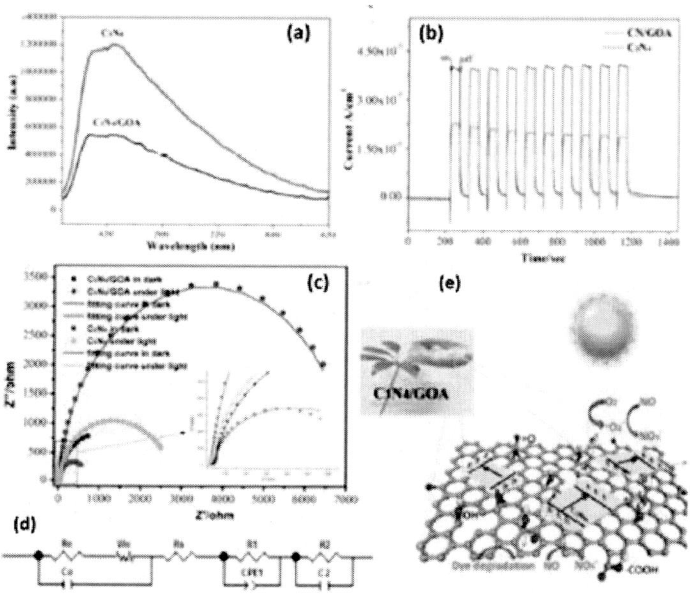

Figure 25. (a) The PL spectra of powder C_3N_4 and C_3N_4/GOA. (b) Photocurrent-time curves of C_3N_4 and C_3N_4/GOA under visible light irradiation. (c) EIS curves of C_3N_4 and C_3N_4/GOA under visible light irradiation and in darkness. (d) The equivalent circuit constructed from these electrodes. (e) The schematic diagram of the photocatalysis mechanism (reproduced with permission from [133]).

Oxidation of NO in ppb level has also been performed over graphene oxide-based aerogels under visible light irradiation. Wan et al., [133] investigated the photocatalytic oxidation of NO over ultra-light C_3N_4/graphene oxide aerogel with density 3-5 mg/cm^3 prepared by one-step cryo-desiccation route (Figure 23).

The maximum NO removal achieved was nearly 50% with 90% wt C_3N_4/GOA under visible light irradiation within 30 min. (Figure 24). PL spectroscopy, photocurrent and electrochemical impedance spectroscopy confirmed the significant separation of photo-induced charge carriers and rapid electronic transport at the interface which resulted in the improved photocatalytic response of the aerogel (Figure 25).

The improved charge separation and rapid charge transfer is attributed to the electronic interaction of C_3N_4 and graphene oxide. The recyclability test confirmed the stability and recyclability of the aerogel (Figure 26).

Yang et al., achieved the photocatalytic degradation efficiency of 51% for degradation of nitric oxide over monolithic WO_3/GO aerogel under visible-light. The interaction between WO_3 and GO resulted in the decrease in rate of recombination of the charge carriers and rapid charge transfer. The recyclability of aerogel was tested for five cycles [134].

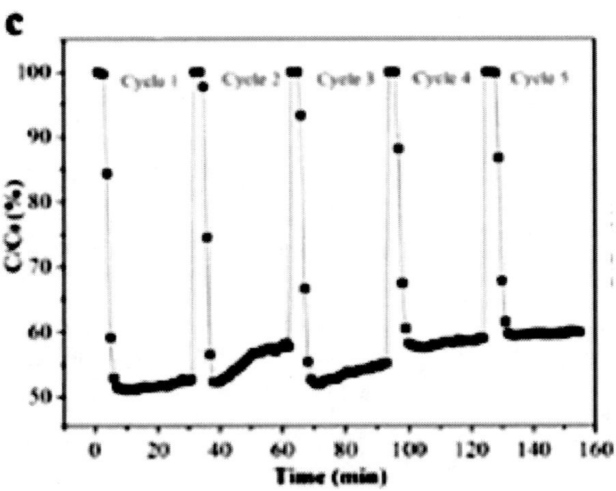

Figure 26. Recyclability test of C_3N_4/GOA for the photocatalytic degradation of NO (reproduced with permission from [133]).

CO_2 Reduction

CO_2 belongs to the category of air pollutants and causes major climate changes. For the last few decades, the level of CO_2 is drastically increasing in the atmosphere due to increasing use of fossil fuels. Therefore, the reduction of CO_2 is essential for environmental safety. In addition, CO_2 can be converted into hydrocarbon fuels [135] by using photocatalytic reduction which is a very efficient method of transforming a harmful greenhouse gas into an energy source with significant industrial applications and address the fossil fuel depletion problems [136-138]. Several catalysts have been used to reduce CO_2 photocatalytically. TiO_2 based aerogels are realized to be effective in the photocatalytic reduction of CO_2. Kreft et al., investigated Cu/TiO_2 aerogel for the photocatalytic reduction of CO_2 to CO in an aqueous solution under UV-A/Vis light in the presence of oxygen gas. Cu(I) was proposed to be the active species for CO_2 photoreduction. Also, high concentration of oxygen resulted in the decrease in reduction rate by suppressing the conversion of Cu(II) to Cu(I) or Cu(0) [139].

Fe^{3+}doped TiO_2 aerogels were synthesized by sol gel process and implemented for the photocatalytic reduction of CO_2 to menthol. The incorporation of Fe^{3+} resulted in the band gap narrowing and slower recombination rate of photo induced charge carriers and hence facilitated in the photoreduction of CO_2 [140].

Similar study for the photocatalytic reduction of CO_2 to menthanol was conducted by Rechberger et al., over TiO_2–Au aerogel monolith under UV light. The trapping of electrons by Au-nanoparticles lead to the suppression of the charge carrier recombination and hence the enhanced photoreduction of CO_2 to methanol with a conversion rate of 2.58 $\mu molg^{-1} h^{-1}$ (Figure 27) [141].

Photocatalytic reduction of CO_2 can also be performed over g-C_3N_4 derived aerogels. Zhang et al., fabricated monolithic Pd-embedded g-C_3N_4/reduced graphene oxide aerogel (Pd-g- C_3N_4/RGOA) by thermal reduction method. The hydrogenation of CO_2 to CH_4 over Pd-g- C_3N_4/RGOA under visible light irradiation resulted in evolution rate of 6.4 $\mu mol\ g^{-1}\ h^{-1}$. Pd facilitated the separation of charge carriers and improvement in photocatalytic response was attributed to the synergic effect of Pd and RGO [142].

3D porous g-C_3N_4/graphene oxide aerogel (CNGA) was synthesized by hydrothermal co-assembly of 2D g-C_3N_4 and graphene oxide (GO) nanosheets (Figure 28) and was studied for the photocatalytic reduction of CO_2 to CO. The efficient photocatalytic response of g- C_3N_4 and 3D networking promoted by GO facilitated the electron transfer [143].

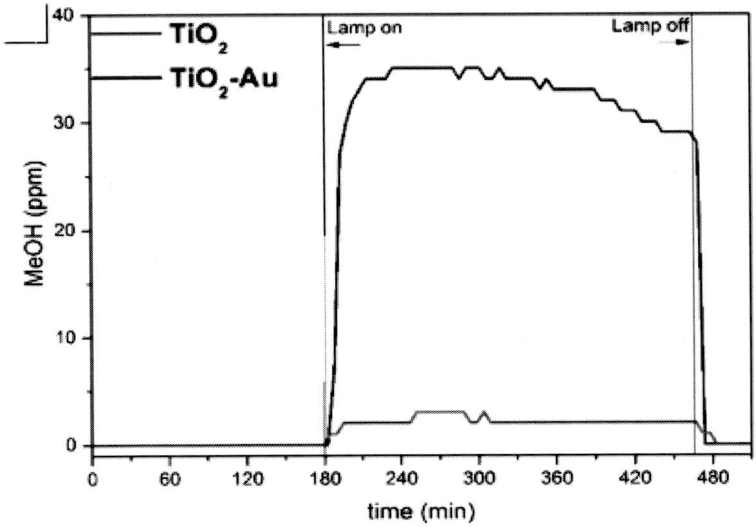

Figure 27. Photocatalytic response of $TiO_2$and TiO_2–Au composite aerogel for reduction of $CO_2$with time and illumination (reproduced with permission from [141]).

Figure 28. Schematic representation for the preparation of g-C_3N_4/graphene oxide aerogel (reproduced with permission from [143]).

Volatile Organic Compounds (VOC)

Besides the photocatalytic reduction of CO_2 and removal of NO_x, aerogels are also found to be potential candidates for the degradation of volatile organic compounds (VOCs) which are one of the most important environmentally hazardous substances like gaseous formaldehyde, benzene, chlorobenzenes etc. Table 3 presents the reports presented by various groups on aerogel based photocatalysts used for the degradation of VOCs.

Table 3. Aerogel photocatalysts used in the photocatalytic degradation of VOCs

Aerogel	Synthesis Route	Pollutant	Photocatalytic response	Reference
$BiVO_4$-quantum tubes decorated rGO-aerogel	Hydrothermal method	Formaldehyde	1.0 ppm to 0.4 ppm in 15 min in the presence of 150W xenon lamp	[144]
titania–silica	Sol-gel process	VOCs	Five fluorescent black lamps (365 nm, Philips 6W) were used as light source	[145]
SiO_2(AG)/ T-ZnO	solvent-exchanging procedure	Nitrobenzene	85% degradation in 150 min Under UV irradiation	[146]
RGO/TiO_2/Ag	Sol-gel method	Formaldehyde	Degradation efficiency of 77.08% within 2 h under UV irradiation	[147]
TiO_2	Purchased	Benzene, toluene, ethylbenzene and oxylene	Efficiency of 24, 67, 84 and 86% using 6W UV lamp irradiation	[148]
TiO_2-SiO_2	sol-gel and ambient pressure drying method	CO, CO_2, HC and NO_x	Degradation rate of 1.3%, 1.5%, 3.2% and 34.4%	[117]
Silica-titania	supercritical drying	Benzene	Complete Degradation to CO_2 under visible light	[149]
Ag/TiO_2	Saturation method	Toluene	95.7% of degradation rate using visible-light lamps	[150]

Future Perspective

The study of aerogels materials depicts that these materials have great potential for various high-performance applications including environment cleaning. However, there are certain unresolved challenges that must be addressed for practical applications to be achievable, as well as some upcoming scientific prospects that appear to be very promising. Photocatalysts' texture and microstructural features have a significant impact on their performance. It is typically helpful to optimize the pore diameters, pore volume, and specific surface area of aerogelsto improve their photocatalytic response. Another efficient technique for achieving these goals is to build hierarchical pore structures. The future investigations can also be more focused on the isotherms and adsorption kinetics of the reactants, as this will be useful in rationally designing the aerogel photocatalysts with greater efficiency.

Additionally, covering various semiconducting materials with some suitable coating (like TiO_2 or C) might be an effective method for photo-oxidative corrosion prevention and hence improving the adsorption capacity. Modification techniques for enhancing the photostability of the photocatalysts, particularly visible light-responsive materials, should be prioritised in future environmental photocatalysis research. As the pores in the aerogels are linked and multi-scale, they can offer effective transit of gas molecules to reaction sitesof the catalyst. In addition to this, the spatial arrangement of the aerogel with the appropriate thickness facilitates the immobilisation and inside collection of the gaseous molecules. This allows for optimum interchange of new gas and product on the aerogel's surface. Despite the fact that aerogels might have a great variety of catalytic applications and have a great influence, they are yet not reach in the market because of their exorbitant process parameters and the prevalence of the barriers such as fragile mechanical strength and low thermal resistance.

Today, such challenges are being reduced to certain level in order to open the way for massive manufactures and uses by employing, for instance, expensive alkoxides are being replaced by low-cost inorganic materials or sustainable biomass generated supplies. Furthermore, ambient pressure drying processes and utilising hybridization chemistry are being adopted for safe aerogel processing to overcome these limitations in the future.

Conclusion

Aerogels are in principle a preferred candidate for use as a medium for the catalytic reduction of environmental pollutant emissions because of their universal synthesis chemistry, sol-gel, along with supercritical drying, that would let material physio-chemistry, textural and morpho-structural properties to be adapted to various catalytic reactions. The steep increase in catalytic activity, selectivity and stability can be ascribed to the coalition of excellent physical properties of the aerogel, such as highly specific surfaces approachable to reactant/product molecules, high dispersion of active species and strong metal or catalyst support or metal-metal interactions. Additionally, because of high and homogeneous dispersion of the catalytically active phase on the aerogel support, the issue of aggregation of nanoscale catalysts due to their high surface-to-volume ratio is alleviated, and hence superior catalytic activity can be attained with an aerogel.

References

[1] McNaught, D., A. Wilkinson, IUPAC Compendium of Chemical Terminology, 2nd ed. (the "Gold Book") Blackwell Scientific Publications, Oxford (1997). (2019) ISBN 0-9678550-9-8.

[2] Kistler, S. S. Nature 127 (1931) 741.

[3] Abdullah H. B., I. Ramli, I. Ismail, N. A. Yusof, *Bull. Mater. Sci.* 42 (2019)241.

[4] Qiu, B., M. Xing, J. Zhang, *J. Am. Chem. Soc.* 136 (2014) 5852.

[5] Ma, H., X. Zheng, X. Luo, Y. Yi, F. Yang, *Materials* 11 (2018) 214.

[6] Higa, K. *Illumin.* 14 (2014) 1.

[7] Shao, X., W. Lu, R. Zhang, F. Pan, *Sci. Rep.* 3 (2013) 3018.

[8] Yang, J., Y. Li, Y. Zheng, Y. Xu, Z. Zheng, X. Chen, W. Liu, *Small* 15 (2019) 1902826.

[9] Korkmaz, S., İ. A. Kariper, *J. Energy Storage*, 27 (2020) 101038.

[10] Yuan, D. S., T. Zhang, Q. Guo, F. X. Qiu, D. Y. Yang, Z. P. Ou, *Chem Eng J.* 351 (2018) 622.

[11] Ashraf, W., T. Fatima, K. Srivastava, M. Khanuja, *Appl. Nanosci.* 9(2019) 1515.

[12] Mittal, H., A. Kumar, M. Khanuja, *J. Saudi Chem. Soc.*23 (2019) 836.

[13] Asharf, W., S. Bansal, V. Singh, S. Barman, M. Khanuja, *RSC Adv.* 10(2020) 25073.

[14] Mittal, H., M. Khanuja, Environ. *Sci. Pollut. Res.* 27 (2020) 23477.

[15] Mittal, H., M. Khanuja, *Dyes Pigm.* 175 (2020) 108109.
[16] Dreyer, M. Titanium dioxide aerogels as photocatalyst for indoor air decontamination, TheUniversity of Oklahoma, Oklahoma, 2002.
[17] Bischoff, L., M. A. Anderson, *Chem. Mater.* 7 (1995) 1772.
[18] Formenti, M., F. Juillet, P. Meriaudeau, *S. J. Teichner, Chem. Tech.* (1971) 680.
[19] Dagan, G., M. Tomkiewicz, *J. Phys. Chem.* A97 (1993) 12651.
[20] Moussaoui, R., K. Elghniji, M. B. Mosbah, E. Elaloui, Y. Moussaoui, *J. Saudi Chem. Soc.* 21 (2017) 751.
[21] Bagheri, S., N. M. Julkapli, *Int. J. Hydrog. Energy* 41 (2016) 14652.
[22] Othmana, I., R. M. Mohamed, F. M. Ibrahem, *J. Photochem. Photobio. A: Chem.* 189 (2007) 80.
[23] Schneider, M., A. Baiker, *J. Mater. Chem.* 2 (1992) 587.
[24] Cao, S., K. L. Yeung, K. C. K. Joseph, M. T. T. Percy, C. T. Y. Samuel, *Appl. Catal. B Environ.* 86 (2009) 127.
[25] Yoda, S., K. Ohtake, Y. Takebayashi, T. Sugeta, T. Sako, T. Sato, *J. Sol-Gel Sci. Technol.* 19 (2000) 719.
[26] Yao, N., S. Cao, K. L. Yeung, *Micropor. Mesopor. Mater.* 117 (2009) 570.
[27] Morales-Torres, S., F. Carrasco-Marín, A. F. Pérez-Cadenas, *F. J. Maldonado-Hódar, Catalysts*, 5 (2015) 774.
[28] Yagub, M. T., T. K. Sen, S. Afroze, H. Ang, *Adv. Colloid Interf. Sci.* 209 (2014) 172.
[29] Pereira, L., M. Alves, Springer, *Dordrecht* (2012) 111.
[30] Jadhav, S. A., H. B. Garud, A. H. Patil, G. D. Patil, C. R. Patil, T. Dongale, P. S. Patil, *Colloids Interface Sci. Commun.* 30 (2019) 100181.
[31] Chung, K. T. *J. Environ. Sci. Health C Environ. Carcinog. Ecotoxicol. Rev.* 34 (2016) 233.
[32] Gogate, P. R., A. B. Pandit, (2004) *Adv. Environ. Res.* 8 (2004) 501.
[33] Hong, S., M. Elimelech, *J. Membr. Sci.* 132 (1997) 159.
[34] Guo, H., K. Lin, Z. Zheng, F. Xiao, S. Li, *Dye Pigm.* 92 (2012) 1278.
[35] Hua, W. K., T. H. Hung, H. Y. Hsu, *Chemosphere* 36 (1998) 2763.
[36] Siddiqui, I., H. Mittal, V. K. Kohli, P. Gautam, M. Ali, M. Khanuja, *Mater. Res. Express,* 5 (2018) 125020.
[37] Mittal, H., M. Khanuja, *Sep. Purif. Technol.* 254 (2020) 117508.
[38] Kumar, S. Singh, M. Khanuja, *Mater. Chem. Phys.* 243 (2020) 122402.
[39] Singh, S., R. Pendurthi, M. Khanuja, S. S. Islam, S. Rajput, S. M. Shivaprasad, *Appl. Phys. A* 123 (2017) 184.
[40] Capelo-Martínez, J. L., P. Ximénez-Embún, Y. Madrid, C. Cámara, *Trac-Trend Anal. Chem.* 23 (2004) 331.
[41] Huang, X., C. Tan, Z. Yin, H. Zhang, *Adv. Mater.* 26 (2014) 2185.
[42] Shi, M., W. Wei, Z. Jiang, H. Han, J. Gao, J. Xie, *RSC Adv.* 6 (2016) 25255.

[43] Papageorgiou, S. K., F. K. Katsaros, E. P. Favvas, G. Em. Romanos, C. P. Athanasekou, K. G. Beltsios, O.I. Tzialla, P. Falaras, *Water Res*., 46 (2012) 1858.

[44] Li, X., S. Yang, J. Sun, P. He, X. Xu, G. Ding, *Carbon* 7 8 (2014) 38.

[45] Li, S., X. Hao, X. Dai, T. Tao, *J. Nanomater*. 2018 (2018) 1.

[46] Zhang, J., S. Fang, J. Mei, G. Zheng, X. Zheng, X. Guana, *Sep. Purif. Technol*. 194 (2018) 96.

[47] Choi, J., D. Reddy, M. Islam, M. Rory, T. K. Kim, *J. Alloy Compd*. 688(2016) 527.

[48] Kim, Y., G. Shao, S. Jeon, S. M. Imran, P. B. Sarawade, H. T. Kim, *Chem. Eng. J.* 231(2013)502.

[49] Xu, Z., J. Jiang, Q. Zhang, G. Chen, L. Zhou, L. Li, *J. Colloid Interf. Sci*. 563 (2020) 131.

[50] Tong, Z., D. Yang, J. Shi, Y. Nan, Y. Sun, Z. Jiang, *ACS Appl. Mater. Interfaces* 7(2015)25693.

[51] Yan, X., R. Xu, J. Guo, X. Cai, D. Chen, L. Huang, Y. Xiong, S. Tan, *Mater. Res*. Bull. 96 (2017) 18.

[52] Huang, X., J. Liu, F. Shi, Y. Ling, S. Liu, *Mater. Des*. 110(2016)624.

[53] Liu, J., X. Wang, F. Shi, Y. Ling, S. Liu, S. Hu, D. Liu, *Adv. Powder Technol*. 27(2016) 1781.

[54] Yu, Y., M. Zhu, W. Liang, S. Rhodes, J. Fang, *RSC Adv*. 5(2015) 72437.

[55] Xu, X., F. Ming, J. Hong, Y. Xie, Z. Wang, *Mater. Lett*. 179(2016)52.

[56] Li, Y., J. Yang, S. Zheng, W. Zeng, N. Zhao, M. Shen, *Ceram. Int.* 42(2016)19091.

[57] Tang, L., C. Jia, Y. Xue, Y. Xue, L. Li, A. Wang, G. Xu, N. Liu, M. Wu, *Appl. Catal. B* 219(2017) 241.

[58] Jiang, Y., S. Chowdhury, R. Balasubramanian, *J. Colloid Interf. Sci*. 534 (2019) 574.

[59] Xiaolin, D., W. Zi, P. Jinjing, G. Wenli, L. Qiao, L. Lin, Y. Juming, *React. Funct. Polym*. 138 (2019) 79.

[60] Saeed, R. M. Y., Z. Bano, J. Sun, F. Wang, N. Ullah, Q. Wang, *J. Appl. Polym. Sci.* 136 (2019) 47404.

[61] Jiang, Y., S. Chowdhury, R. Balasubramanian, *Environ. Pollut*. 251 (2019) 344.

[62] Najafidoust, A., M. Haghighi, E. A. Asl, H. Bananifard, *Sep. Purif. Technol*. 221 (2019) 101.

[63] Liu, X., J. Wang, Y. Dong, H. Li, Y. Xia, H. Wang, *Mater. Sci. Semicond.* Process. 88 (2018) 214.

[64] Li, Z., H. Wang, X. Wei, X. Liu, Y. Yang, W. Jiang, *J. Alloys Compd*. 659 (2016) 240.

[65] Liu, L., P. Zhao, S. Wang, N. Xie, H. Wang, V. Strokova, L. Lu, X. *Cheng, Compos. Interfaces*, (2020) 1–17. doi: 10.1080/09276440.2020.1789932.
[66] Sud, D., G. Mahajan, M. P. Kaur, *Bioresour. Technol.* 99 (2008) 6017.
[67] Ngah W. S. W. and M. A. K. M. Hanafiah, *Bioresour. Technol.*, 2008,99, 3935.
[68] Zhang, R., W. Wan, D. Li, F. Dong, Y. Zhou, *Chinese J. Catal.* 38 (2017) 313.
[69] Yang, M. Q., N. Zhang, Y. Wang, Y. J. Xu, *J. Catal.* 346 (2017) 21.
[70] Reddy, A., J. Choi, S. Lee, R. Ma, T. K. Kim, *RSC Adv.* 5(2015) 18342.
[71] Xiao, W., W. Zhou, Y. Zhang, L. Tian, H. Liu, Y. Pu, *J. Nanomater.* 2016 (2016) 1.
[72] Mompelat, S., B. LeBot, O. Thomas, *Environ. Int.* 35 (2009) 803.
[73] Wang, J., Y. Lin, L. Chen, *Analyst*118 (1993) 277.
[74] Kasiotis, K. M., C. Emmanouil, P. Anastasiadou, A. Papadi-Psyllou, A. Papadopoulos, O. Okay, K. Machera, *Chemosphere* 119 (2015) S145.
[75] Groz, M. P., M. J. M. Bueno, D. Rosain, H. Fenet, C. Casellas, C. Pereira, V. Maria, M. J. Bebianno, E. Gomez, *Sci. Total Environ.* 493 (2014) 162.
[76] Kolpin, W., E. T. Furlong, M. T. Meyer, E. M. Thurman, S. D. Zaugg, L. B. Barber, H. T. Buxton, *Environ. Sci. Technol.* 36 (2002) 1202.
[77] White, R., de Groot AC (2006) Cosmetics and skin care products. In: Frosch P, Menné T, Lepoitevin J (eds) *Contact dermatitis*. Springer, Berlin-Heidelberg, pp 493–506.
[78] Ramirez, J., R. A. Brain, S. Usenko, M. A. Mottaleb, J. G. O'Donnell, L. L. Stahl, J. B. Wathen, B. D. Snyder, J. L. Pitt, P. Perez-Hurtado, *Environ. Toxicol. Chem.* 28 (2009) 2587.
[79] Rivera-Utrilla, J., M. Sánchez-Polo, M. A. Ferro-García, G. Prados-Joya, R. Ocampo-Pérez, *Chemosphere* 93 (2013) 1268.
[80] Tambosi, J. L., L. Y. Yamanaka, H. J. José, R. F. P. M. Moreira, H. F. Schröder, *Quim Nova* 33 (2010) 411.
[81] Carballa, M., F. Omil, J. M. Lema, M. Llompart, C. García-Jares, I. Rodríguez, M. Gómez, T. Ternes, *Water Res.* 38 (2004) 2918.
[82] Pap, Z., A. Radu, I. J. Hidi, G. Melinte, L. Diamandescu, T. Popescu, L. Baia, V. Danciu, M. Baia, Chinese *J. Catal.* 34 (2013) 734.
[83] Sadrieyeh, S., R. Malekfar, *J. Non-Cryst. Solids* 489 (2018) 33.
[84] Liu, M., D. Zou, T. Ma, Z. Liu, Y. Li, *Inorg. Chem. Front.*, 6 (2019) 1388.
[85] Bolobajev, J., M. Kask, K. Kreek, M. Kulp, M. Koel, A. Goi, *Chem. Eng. J.* 357 (2019)120.
[86] Liu, Y., D. Yang, Y. Shi, R. Yu, L. Song, J. Qu, Z. Yu, *ACS Sustainable Chem. Eng.* 7 (2019) 11228.
[87] Zhao, X., X. Wang, J. Zhang, X. Yi, *Chem. Asian J.* 10.1002/asia.201801561.

[88] Zhanga, J., Z. Wang, M. Fan, P. Tong, J. Sun, S. Dong, J. Sun, *J. Mater. Res. Technol.* 8 (2019) 4577.

[89] Li, Y., Y. Yang, P. Yang, L. Jiang, W. Wang, J. He, Y. Chen, *J. Wang, Catal. Commun.* 145 (2020) 106121.

[90] Tan, L., C. Yu, M. Wang, S. Zhang, J. Sun, S. Dong, J. Sun, *Appl. Surf. Sci.* 467–468 (2019)286.

[91] Prasad, S., P. Shanmugam, K. Bhuvaneswari, G. Palanisamy, T. Pazhanivel, T. Arunkumar, M. S. AlSalhi, M. J. *Aljaafreh, Crystals* 10 (2020) 300.

[92] Chen, S., W. Lu, J. Han, H. Zhong, T. Xu, G. Wang, W. Chen, *Chem. Eng. J.* 359 (2019)119.

[93] Saccoa, O., V. Vaiano, C. Daniel, W. Navarraa, V. Venditto, *Mater. Sci. Semicond. Process.* 80 (2018) 104.

[94] Z. Deng, J. Wang, Y. Zhang, Z. Weng, Z. Zhang, B. Zhou, J. Shen, L. Cheng, *Nanostructured Materials* 11 (1999) 1313.

[95] Dagan, M. Tomkiewicz, *J. Non-Crystal. Solids* 175 (1994) 294.

[96] X. Zhao, J. Zhang, X. Wang, J. Zhang, B. Liu, X. Yi, *Appl. Surf. Sci.* 478 (2019) 266.

[97] I. Lázár, J. Kalmár, A. Peter, A. Szilágyia, E. Gyori, T. Ditrói, I. Fábián, *Appl. Surf. Sci.* 356 (2015) 521.

[98] Chen, L., J. Zhu, Y. Liu, Y. Cao, H. Li, H. He, W. Dai, K. Fan, *J Mol. Catal. A Chem.* 255 (2006) 260.

[99] Orellana-García, F., M. A. Álvarez, M. V. López-Ramón, J. Rivera-Utrilla, M. Sánchez-Polo, M. Á. Fontecha-Cámara, *Appl. Catal. B Environ.* 181 (2016) 94.

[100] Nawaz, M., W. Miran, J. Jang, D. S. Lee, *Appl. Catal. B Environ.* 203 (2017) 85.

[101] Tomkiewicz, M., G. Dagan, Z. Zhu, *Res. Chem. lntermed.* 20 (1994) 701.

[102] Liang, S., Y. Zhou, K. Kang, Y. Zhang, Z. Cai, J. Pan, *J. Photochem. Photobiol. A* 346 (2017) 1.

[103] Han, X., L. P. Naeher, *Environ. Pollut.* 32 (2006)106.

[104] Manisalidis, I., E. Stavropoulou, A. Stavropoulos, E. Bezirtzoglou, *Front. Public Health*, 8(2020) 14.

[105] Sweileh, W. M., S. W. Al-Jabi, S. H. Zyoud, A. F. Sawalha, *Multidiscip. Respir. Med.* 13 (2018) 15.

[106] Goyal, P., Sidhartha. *Atmos. Environ.*37 (2003) 5423.

[107] Stanek, L. W., J. S. Brown, Reference Module in Biomedical Sciences https://doi.org/10.1016/B978-0-12-801238-3.11384-4.

[108] Montero-Montoya, R., R. López-Vargas, O. Arellano-Aguilar, *Ann. Glob. Health* 84 (2018)225.

[109] Lian, Z. H., F. Liu, H. He, X. Y. Shi, J. S. Mo, Z. B. Wu, *Chem. Eng. J.* 250 (2014) 390.

[110] Noris, M., M. Todeschini, F. Casiraghi, D. Roccatello, G. Martina, L. Minetti, B. Imberti, F. Gaspari, M. Atti, G. Remuzzi, *Am. J. Kidney. Dis.* 32 (1998) 115.

[111] Barnes, J. M., W. A. Apel, K. B. Barrett, *J. Hazard. Mater.* 41 (1995) 315.

[112] Kim, K. J., H. G. Ahn, *Microporous Mesoporous Mater.* 152 (2012) 78.

[113] Xiao, B., P. S. Wheatley, X. B. Zhao, A. J. Fletcher, S. Fox, A. G. Rossi, I. L. Megson, S. Bordiga, L. Regli, K. M. Thomas, R. E. Morris, *J. Am. Chem. Soc.* 129(2007) 1203.

[114] Brown, C. E., P. G. Hall, J. Colloid. *Interface. Sci.*, 1973, 42, 334–341.

[115] Cooper, C. D. Kirk-OthmerEncyclopedia of Chemical Technology, Air Pollution *Control Methods*26(2000)667.

[116] Wu, Z. S., S. Yang, Y. Sun, K. Parvez, X. Feng, K. Müllen, *J. Am. Chem. Soc.* 134 (2012) 9082e9085.

[117] Zhang, S., Z. Zhang, J. Pei, R. Li, J. Zhang, J. Cai1, J. Cui, *Mater. Res. Express* 5 (2018)025036.

[118] Zhou, Y., W. Li, Q. Zhang, S. Yan, Y. H. Cao, F. Dong, F. Wang, *Phys. Chem. Chem. Phys.* 19 (2017) 25610.

[119] Ma, J. Z., C. X. Wang, H. He, *Appl. Catal. B* 184 (2016) 28.

[120] Doronkin, D. E., A. B. Kuriganova, I. N. Leontyev, S. Baier, H. Lichtenberg, N. V. Smirnova, J. D. Grunwaldt, *Catal. Lett.* 146 (2016) 452.

[121] Olsson, L., H. Persson, E. Fridell, M. Skoglundh, B. Andersson, *J. Phys. Chem.* B105 (2001) 6895.

[122] Zhou, Y., X. J. Zhang, Q. Zhang, F. Dong, F. Wang, Z. Xiong, *J. Mater. Chem. A* 2 (2014)16623.

[123] Liu, Y., S. Yu, Z. Y. Zhao, F. Dong, X. A. Dong, Y. Zhou, *J. Phys. Chem.* C, 121 (2017)12168.

[124] Xiong, T., M. Q. Wen, F. Dong, J. Y. Yu, L. L. Han, B. Lei, Y. X. Zhang, X. S. Tong, Z. G. Zang, *Appl. Catal. B* 199 (2016) 87.

[125] Cai, S. Y., S. Yu, W. C. Wan, W. Wen, Y. Zhou, *RSC Adv.* 7 (2017) 27397.

[126] Ouyang, S. X., P. Li, H. Xu, H. Tong, L. Q. Liu, J. H. Ye, *ACS Appl. Mater.* Interfaces 6(2014) 22726.

[127] Onrubia, J. A., B. Pereda-Ayo, U. De-La-Torre, J. R. González-Velasco, *Appl. Catal. B*213 (2017) 198.

[128] Wang, Z. Y., Y. Huang, W. K. Ho, J. J. Cao, Z. X. Shen, S. C. Lee, *Appl. Catal. B* 199(2016) 123.

[129] Luo, J. M., G. H. Dong, Y. Q. Zhu, Z. Yang, C. Y. Wang, *Appl. Catal. B* 214(2017) 46.

[130] Hu, J., D. Chen, N. Li, Q. Xu, H. Li, J. He, J. Lu, *Small* 14 (2018) 1800416.

[131] Hu, J., D. Chen, N. Li, Q. Xu, H. Li, J. He, J. Lu, *Appl. Catal. B Environ.* 236 (2018) 45.

[132] Liu, D., D. Chen, N. Li, Q. Xu, H. Li, J. He, J. Lu, *Appl. Catal. B Environ.* 243 (2019) 576.

[133] Wan, W., S. Yu, F. Dong, Q. Zhang, Y. Zhou, *J. Mater. Chem. A*, 4 (2016) 7823.

[134] Yang, L., Y. Liu, R. Zhang, W. Li, P. Li, X. Wang, Y. Zhou, *Chinese J. Catal.* 39 (2018) 646.

[135] Kočí, K., L. Obalová, and Z. Lacný, *Chem. Pap.* 62 (2008) 1.

[136] McDaniel, N. D., S. Bernhard, *Dalton Trans.* 39 (2010) 10021.

[137] Habisreutinger, S. N., L. Schmidt-Mende, J. K. Stolarczyk, *Angew. Chem., Int.* Ed. 52. (2013) 7372.

[138] Xu, B. Q., J. M, Wei, H. Y. Wang, K. Q. Sun, Q. M. Zhu, *Catal. Today* 68 (2001) 217.

[139] Kreft, S., R. Schoch, J. Schneidewind, J. Rabeah, E. V. Kondratenko, V. A. Kondratenko, H. Junge, M. Bauer, S. Wohlrab, *M. Beller, Chem* 5 (2019) 1818.

[140] Chen, X., X. Ye, J. He, L. Pan, S. Xu, C.Xiong, *J. Sol-Gel Sci. Technol.*95 (2020) 353.

[141] Rechberger, F., M. Niederberger, *Mater. Horiz.* 4 (2017) 1115.

[142] Zhanga, R., Z. Huang, C. Li, Y. Zuo, Y. Zhou, *Appl. Surf. Sci.* 475 (2019) 953.

[143] Tong, Z., D. Yang, J. Shi, Y. Nan, Y. Sun, Z. Jiang, *ACS Appl. Mater.* Interfaces 7 (2015)25693.

[144] Jianlong, Y., Q. Shi, R. Zhang, M. Xie, X. Jiang, F. Wang, X. Cheng, W. Han, *Carbon* 138 (2018) 118.

[145] Shengli, C., K. L. Yeung, P. Yue, *Appl. Catal. B Environ.* 68(2006) 99.

[146] Yi, Z., T. Jiang, Y. Cheng, Q. Tang, *Nanotechnol. Rev.* 9 (2020) 1009.

[147] Wang, G. Wang, Y. Zhang, Y. Ma, Z. Wu, D. Gao, R. Yang, B. Wang, X. Qi, *J. Yang, Sci. Rep.* 9 (2019)1.

[148] Ao, C. H., S. C. Lee, C. L. Mak, L. Y. Chan, *Appl. Catal. B Environ.* 42 (2003) 119.

[149] Yoda, S., D. J. Suh, T. Sato, *J. Sol-Gel Sci. Technol.* 22 (2001) 75.

[150] Jafari, J., M. Kermani, A. Hosseini-Bandegharaei, A. Rastegar, M. Gholami, A. Alahabadi, G. Farzi, *Chem. Eng. Res. Des.* 150 (2019) 1.

[151] Wang, Y., Y. Z. Zheng, S. Lu, X. Tao, Y. Che, *J. F. Chen, Appl. Mater. Inter.*, 7 (2015)6093–6101.

[152] He, D., Y. L. Li, I. S. Wang, J. S. Wu, Y. L. Yang, *Q. E. An Applied Surface Science,* 391(2017a) 318–325.

[153] Shi, W. L., F. Guo, H. B. Wang, M. M. Han, H. Li, S. L. Yuan, H. Huang, Y. Liu, Z. H. Kang, *Applied Catalysis B: Environmental*, 219 (2017) 36–44.

Chapter 4

Overview of Aerogels Applications Aimed at Thermal and Acoustic Insulation

Norma Gisela Corral Capulin*
Dirección de Ingeniería en Biotecnología, Universidad Politécnica del Valle de Toluca, Almoloya de Juárez, Estado de México

Abstract

Materials science and engineering are fields of knowledge whose efforts result the development of new materials, devices and technologies. The new materials can be used in the improvement of processes increasing their efficiency and reducing production costs. In this context, aerogels are versatile materials due to their unique properties such as high porosity (> 90%), fine pore size (<100 nm), large specific surface (500-100h m^2 g^{-1}), and low thermal conductivity. Aerogels applications are growing in different sectors including construction, cryogenics and aerospace. The present work provides an overview of their applications mainly as thermal and acoustic insulators.

Keywords: aerogels, insulation, application

Introduction

The scientific story of aerogels started almost a century ago in 1931 with Samuel Stephens Klister at the College of the Pacific in Stockton, California. Stephens proposed to remove all the liquid in a jelly without any shrinkage

* Corresponding Author's Email: gisela.corral@upvt.edu.mx.

In: Aerogels: Properties and Applications
Editor: Alina Iuliana Pruna
ISBN: 978-1-68507-788-4

which lead to the origin of a new class of materials called aerogels. These materials present low density, large specific surface areas and can be made for almost any material [1]. Klister's contribution to aerogels design parts from the replacement of liquid in the pores of a gel with air resulting in a moderate shrinkage of the matrix. The term aerogels describe materials synthesized by sol-gel chemistry in which the liquid portion of the gel is replaced with gas by a proper drying technology (Figure 1).

Hüsing and Shubert [2] gave the most commonly accepted definition of aerogels as "the materials in which the typical pore structure and network are largely maintained when the pore liquid of a gel is replaced by air." Recently Leventis [3] stated that aerogels are an open non-fluid colloidal network or polymer network that is expanded throughout its whole volume by gas and is formed by the removal of all swelling agents from a gel without substantial volume reduction or network compaction. Nevertheless, the development of synthesis methods, characterization, and applications of aerogels by scientists keep generating definitions of aerogels. Ziegler et al. [1] define aerogel as a solid with meso- and macropores with diameters up to a few hundred nanometers and a porosity of more than 95% in which the dispersed phase is a gas [4]. Therefore, the term aerogel describes a special geometric structure of matter that is neither limited to any material or any synthesis method. Hence the possible variety of materials and therefore the multitude of their applications are almost unbounded.

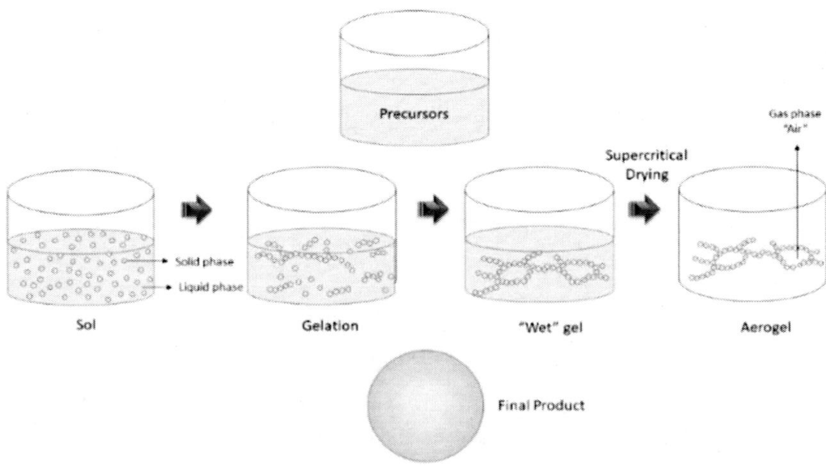

Figure 1. The general process for aerogels synthesis by sol-gel method.

Properties and General Applications

Physical properties (optical, thermal, acoustic, electrical, and mechanical) of aerogels are originated from the combination of a solid network and nanosized pores filled with air. The choice of the precursors (Silica-based aerogels, organic-inorganic silica aerogels, organic aerogels, and bioaerogels) and the optimization of sol-gel parameters determine the physical properties of the final aerogel products and therefore, their possible applications. Figure 2 summarizes the applications of aerogels based on their physical characteristics [2, 5].

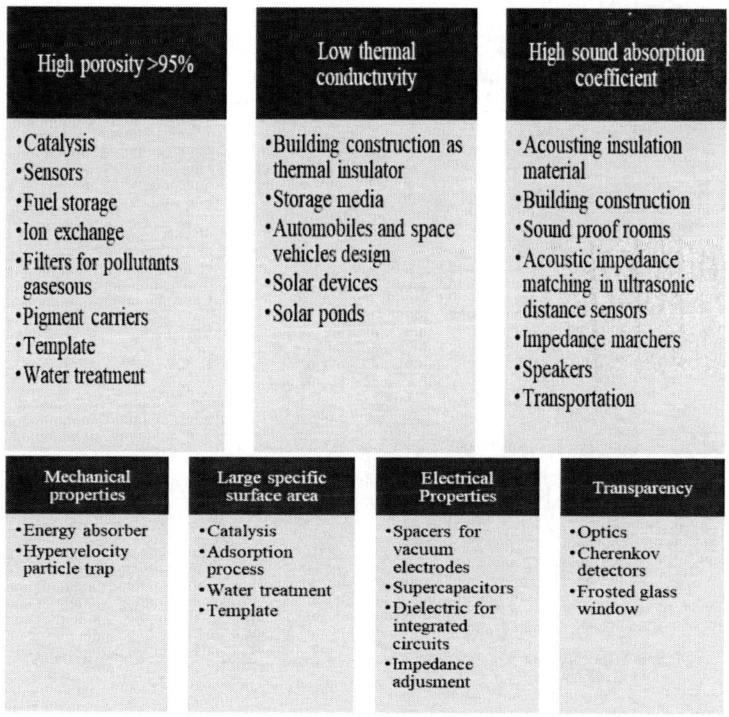

Figure 2. Applications according to aerogels physical properties.

Also, aerogels applications can be divided into different fields. The most important applications for aerogels are in their use as sensors, thermal insulation, acoustic insulation, waste management, molten metals, optics and light guides, electronic devices, capacitors, imaging devices, catalysts, pesticides, and cosmic dust collection [5]. Therefore, they acquire great relevance in different fields of science and technology as shown in Figure 3.

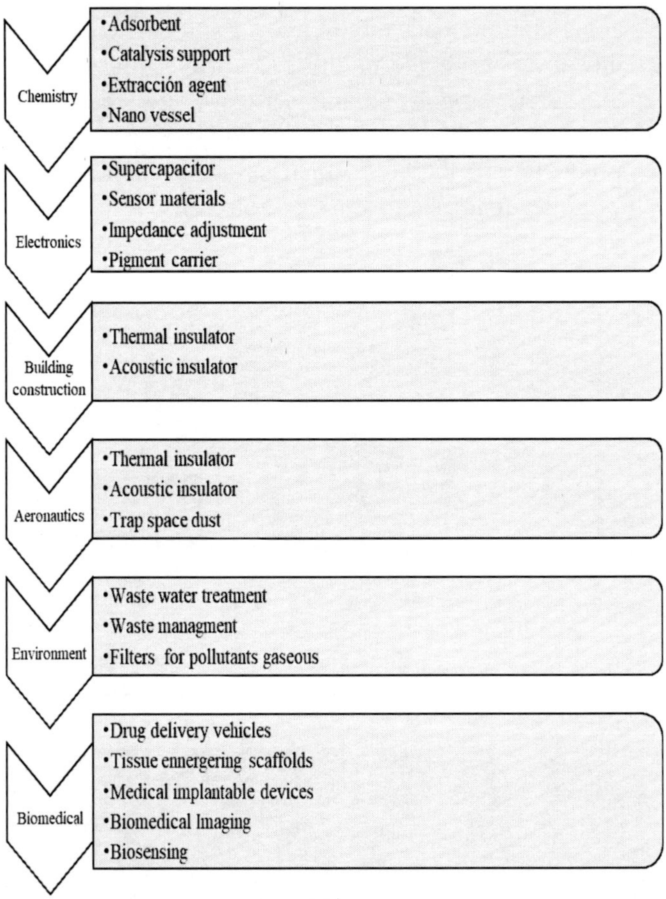

Figure 3. Aerogels applications in different fields of science and technology.

Other Applications

Figures 2 and 3 summarize key applications of aerogels in different fields according to their features. However, their use is not limited to these sectors. In the case of the agriculture industry, aerogels have been used as a potential carrier of chemical compounds such as herbicides, pesticides, and fertilizers. Due to fine small size and large surface area can also act as a pesticide removing the protecting lipid layer of insects by abrasion or adsorption to lose body fluid eventually causing their death. Moreover, silica aerogel powders have been applied for the storage protection of grains [2, 6].

In the space industry, the National Aeronautics and Space Administration (NASA) used aerogels as standard media for collecting cosmic dust aboard Stardust spacecraft taking advantage of their extreme lightness and unique transparent mesostructured. This enables a soft landing of very fast extraterrestrial particles impinging with a velocity of more than 3 km/s [2]. The particle vaporizes on impact with solids and passes through gases, which also can be trapped in aerogels. NASA also employed aerogels for thermal insulation of the Mars Rover and space suits. Moreover, US Navy is evaluating aerogel undergarments as passive thermal protection for divers [5]. Low-density silica aerogels have been used to study the superfluid transition of ^4He and phase separation of ^3He-^4He mixtures [2].

In the environment field, aerogels application involves air cleaning, particularly CO_2 and Volatile Organic Compounds (VOC), such as benzene, toluene, ethylbenzene, and xylene which are also now as BTEX. The adverse effects of air pollutants on the environment and humans make a priority their removal. In the case of CO_2, it is one of the primary greenhouse gasses in the atmosphere, while the VOC are extremely carcinogenic. In the removal of air pollutants, aerogels can be used as sorbent media or as support for adsorbents, taking into account properties such as selectivity, hydrophobicity, high porosity, and specific surface area. Such properties can be modified through the incorporation of organic functional groups chemically related to the species to be eliminated. The adsorption capacity can be increased as well during the synthesis. Modified aerogels to support amine groups allow the capture of CO_2 gas.

Synthesis techniques such as sol-gel create microstructural patterns, control the composition and surface chemistry of aerogel. The result is an aerogel capable to capture CO_2. Therefore, recent research targets CO_2 capture materials by manipulating the chemophysical properties of different aerogels. In this regard, amine-modified SiO_2 aerogels using 3-aminopropyltriethoxysilane (APTES), polyethyleneimine (PEI), and tetraethylenepentamine (TEPA) have been explored with CO_2 adsorption capacity of 6.97 mmol/g (in 10% $CO_2 + N_2$) and 7.9 mmol/g under simulated flue gas conditions (in prehumidified 10% CO_2), respectively [6].

The application of aerogels in water treatment derives from the porosity of aerogels. They are of significant interest for the separation of pollutants from water particularly oil, dyes, toxic organic compounds, and heavy metal ions like chromium, lead, mercury, copper, cadmium, and cobalt. The discharge of waste causes serious contamination of aquatic ecosystems [7]. Li et al. [8] synthesized porous and shape-adjustable magnetic aerogel based on Poplar Catkins (PCs), chitosan (C), and Fe_3O_4 (M) (PCCM) for the adsorption of Cu(II) and Pb(II) from wastewater to increase affinity for metal ions, surface area, and mechanical properties. According to the Langmuir isotherm, the maximum adsorption capacity of Cu(II) and Pb(II) were 81.3 and 112.3 mg g^{-1}, respectively. The efficient adsorption capacity was attributed to the Cu(II) and Pb(II) chelated with amine and O-containing groups. The PCCM aerogel was able to be separated easily from wastewater with an external magnetic field and used for seven cycles. The result demonstrated that the PCCM aerogel could be applied as a potential adsorbent of Cu(II) and Pb(II) in wastewater [6].

The introduction of organic aerogels and bioaerogels is relatively recent. These biomaterials exhibit a large porosity, controlled pore diameter, large internal surface, interconnected 3D structure, together with biodegradability, and proven biocompatibility. The aerogels attract great interest in the biotechnology community due to their potential applications, such as ultrasound contrast agents, biomedical implantable devices, tissue engineering, disease diagnosis, antibacterial materials, drug delivery, and biosensing. For instance, the high mesoporosity in aerogels which is analogous to the native extracellular matrix (>80% porosity), makes them an ideal biomaterial in regenerative medicine. Aerogels can also meet the performance requirements for other emerging life science and personal care fields, such as cosmetics, food packaging, and the pharmaceutical industry sector [9].

In this context, aerogels produced from marine polymers, such as chitosan and alginate are of interest for wound healing applications due to their attractive properties like high porosity along with the antimicrobial activity of chitosan or the capacity to provide a moist environment of alginate. Nowadays the synthesis of hybrid materials is reported like the obtention of alginate-chitosan aerogel by Batista et al. [10]. The resulting aerogels were observed to be light-weight, fluffy mesoporous fibers with a specific surface area of 162–302 m^2 g^{-1} and a specific pore volume of 1.41–2.49 cm^3 g^{-1}. The biocompatibility of the fibers was evaluated resulting in not being cytotoxic. Bioactivity of the fibers regarding the ability to close a wound on an in vitro scale and antibacterial activity were also evaluated. Aerogel fibers presented percentages of recovered scratch area of about 75%, higher than the untreated control (~50%), and clear antibacterial activity against *Staphylococcus aureus* and *Klebsiella pneumoniae*. The obtained results suggest that these alginate-chitosan aerogel fibers could be good candidates for wound healing applications after comparing with a commercial medical device (Kaltostat ®) and revelated a similar performance in the cell experiments.

Aerogels also can be utilized as a scaffold for encapsulation and immobilization of various biomolecules for a variety of applications. For instance, the in-situ encapsulation of biomolecules such as proteins and enzyme catalysts into the nanostructure of aerogels can significantly improve their performance and biocatalytic activity. The biomolecules could be incorporated into aerogels either after supercritical drying or can be encapsulated in the wet gel during sol-gel reaction followed by the supercritical drying process. Lipase is one of the most studied enzymatic catalysts that could be incorporated into aerogel nanostructure during a sol-gel reaction without being affected by gelation and harsh supercritical drying condition. In most cases, the catalytic activity of lipase remained intact or even could be improved when encapsulated in an aerogel matrix.

Also, aerogels contain promising applications in food processing and preservation technologies. Da Silva et al. [11] generate cross-linked aerogels by using germinated and non-geminated wheat starch, the resulting aerogels exhibited physical integrity, high water adsorption capacity, and high degradation temperature. Thus, they can be used as excellent absorbents for food packages, food microencapsulation, and food carrier materials. Taking advantage of aerogel's excellent thermal insulation properties, polysaccharide-based aerogels might be good competitors for current polystyrene packaging (used to retain the desired temperature of packaged chilled or hot food).

The specific functionality could act as an antibacterial platform against different bacteria strains. In this case, the antibacterial activity of chitosan-derived aerogels toward *E. Coli* bacteria has been investigated and confirmed.

Aerogels' applications as a filter in extracorporeal devices are possible owing to its high porosity and anti-fouling properties as well as the ability to be functionalized. Some aerogel formulations have found applicability for being used as biocompatible bone graft and dental substitutes.

In function with the last information, aerogels have a great number of applications in different fields of science and technology. Nowadays the main use for this kind of material is thermal and acoustic insulation. According to the report by Grand View Research, the global aerogel market size was estimated at $538.14 million USD in 2016 and is projected to expand at a CARG of 22.6% during the forecast period up to 2025. Factors such as efficient thermal and sound resistance as well as lighter and thinner alternatives to convectional insulations are major growth drives for the market [12].

It is expected that the aerogel markets will continue to grow much more rapidly than the "conventional" insulation business for at least a decade up to the point when markets will begin to saturate. Saturation will certainly depend on a continuous reduction of product cost which goes hand in hand with increasing production capacities. The value for a cubic meter of silica aerogel is on the order of $4,000 USD. With increasing commercialization, this value could drop below the $1,500 USD mark by the year 2021 [13]. Industries like oil and gas, building and construction, automotive, aerospace, and marine are prominent end-users. Among them, the building and construction industry is expected to be one of the key end-use segments in the market, especially in developed regions. Buildings consume a large amount of energy; for instance, 16.6% of the overall energy in the U.S. is used in heating and cooling buildings. Therefore, the improvements of thermal insulation systems have prospects of large savings in primary energy consumption. It can be concluded that aerogels have great potential in a wide range of applications as energy-efficient insulation, windows, acoustics, and so forth [12, 14].

With a large number of synthetic methods available to fabricate insulating aerogels are many products that are available commercially today. Those are almost exclusively based on silica. The first commercial aerogels were based on SiO_2, following a recipe similar to Kistler's original procedure, and were produced after 1940 by the Monsanto Chemical Corp., Everett, Massachusetts, USA. Production stopped in the sixties for more or less two decades but started again in the early eighties and during the nineties. Different types of aerogels were then produced for various applications but a large-scale commercial adaptation by the thermal insulation trade took place only recently. At present, the three main players involved in this field are the North American industrials Cabot Corporation, Aspen Aerogels, and NanoPore. The products by these companies are available in the form of granular material or flexible blankets. In Table 1, mainly commercial aerogel products and their applications in different fields are enlisted.

Table 1. Commercial aerogel products

Company	Products	Thermal conductivity W m^{-1} K^{-1}	Applications
Cabot Corporation	Lumira® aerogel Enova ® aerogel	- -	Interior and exterior insulating plasters for breathable building envelopes and facades. Insulation boards for internal insulating finishing systems. Thermal insulation coatings for safe-to-touch surfaces, energy efficiency, prevention of corrosion under insulation (CUI), thermal breaks, and condensation control. Insulation packs for oil and gas subsea pipelines. Non-wovens for architectural membrane roofing. Industrial insulation. Ultra-low gloss matte paint coatings for industrial surfaces Outdoor gear and apparel. Personal products including skin and beauty care. [15]
Aspen Aerogels	Pyrogel® HPS	0.022	Optimized for high-temperature hydrocarbon & chemical processing, steam pipes, vessels and equipment, gas and steam turbines, removable blankets.
	Pyrogel® XTE	0.022	CUI defense, medium to high-temperature processes, pipes, vessels and equipment, District Energy steam networks, aerospace and defense system
	Pyrogel®XTF	0.022	Passive pool fire and passive jet fire protection for industrial and commercial applications, relief system sizing.
	Cryogel®Z	0.017	Sub-ambient and cryogenic pipelines, vessels and equipment, gas liquefaction & re-gasification facilities.
	Space loft® Subsea	0.0145	Superior flow assurance and space-saving subsea pipe-in-pipe system, electrically trace heated pipe-in-pipe systems.
	Spaceloft	0.0165	Space-saving wall, proof, and roof insulation in residential and commercial buildings.
	Cryogel®x201	0.017	Sub-ambient including refrigerated appliances, cold storage, and aerospace. [16]
NanoPore Incorporated	NanoPore™ Thermal insulation	-	Thermal isolation of microelectronics to refrigeration to high temperature, high-performance exhaust systems [17].

Thermal and Acoustic Insulations

Continuous changes in the environmental landscape, rapid urbanization, industrialization, and health problems such as hearing loss, stress, insomnia, cardiac dysfunction, and high blood pressure can be associated with noise pollution from road traffic, construction sites, and social activities, affecting the well-being of millions of people daily. One way to minimize the harmful effects is using sound insulation, materials absorbing the sound energy during a noise propagation process. In general, sound insulators can be categorized as resonant and porous insulation materials. Resonant materials exhibit sufficient ability to absorb sound close to the resonance frequency, however, their performance is restricted by their narrow absorption frequency bands and their heavyweights. On the other hand, porous sound absorption materials made from fibers possess the advantages of broadband sound absorption, flexibility in structural design, and ease of installation and replacement; hence, they are widely used to eliminate noise in the fields of building and transportation [18, 19]. However, the large pore diameters of commonly used sound absorption materials greatly limit their functions in the viscous-thermal consumption of acoustic energy. To ensure a high sound absorption coefficient for a material, more fibers are needed to achieve the desired sound absorption properties, which leads to an increase in weight and thickness. Therefore, it remains challenging to decrease the weight of commonly used fibrous sound absorption materials while retaining high sound absorption coefficients. In this context nanofibers, aerogels have been found an important field of the application taking into consideration their characteristics such as hierarchical pore structures, large specific surface areas, low density, and high sound absorption coefficient that can guarantee sound waves cannot so easily cross through the material. However, low-density nanofibrous aerogels have some disadvantages due to their fragile bendability and poor tensile strength, based on the above information Cao et al. [18] reported a method to obtain nanofibrous aerogel from bamboo by freeze-drying with robust bendability and superelastic properties. The reinforced nanofibrous aerogel (RNFA's) obtained can maintain their structural integrity after 100 bending-recovery cycles at the beginning angle from 100°, they can also completely revive from large deformation with a fast recovery speed (~834 mm s^{-1}). Moreover, the internal hierarchical porous structures and hydrophobicity of the materials provide densities less than 11 mg cm^{-3}, noise reduction coefficient of 0.41, and good moisture resistance; thus, they are promising options in cabins, vehicles, and the regulation of indoor reverberation.

Ba Thai et al. [19] synthesized a rubber aerogel from tire waste, these aerogels present robust mechanical performance, the density of 91 mg cm^{-3}, highly porous structure (porosity > 90%), high sound absorption efficiency (noise reduction coefficient of 0.56), and low thermal conductivity (0.035–0.049 W m^{-1} K^{-1}). These advanced rubber aerogels can be a promising candidate for thermal and acoustic insulation in cabins, vehicles, buildings, and aerospace. Do et al. [20] synthesized a pineapple aerogel with a Noise Coefficient Reduction of 0.52 with a concentration of pineapple fiber of 2.0 wt.%.

Monolithic aerogel panes with different thicknesses, prepared via a rapid supercritical extraction process, were investigated by Merli et al. [21] in terms of acoustic absorption and transmission loss employing measurements in a Kundt's tube apparatus. Good acoustic performance was achieved: 12.7-mm-thick monolithic panes are characterized by a peak value of the absorption coefficient of 0.88 at about 1500 Hz frequency; when the thickness increases the acoustic absorption coefficient (α) decreases (a range 0.78–0.54), with peaks at lower frequencies (1300 and 1100 Hz). Acoustic absorption performance was compared to that of granular aerogel layers with different granule sizes. The 12.7 mm monolithic pane is characterized by a normal incidence absorption coefficient similar to a layer of large granules; the small granules have the best acoustic absorption performance (α = 1), with NRC-values about two times the monolithic ones at the same thickness. Moreover, NRC values are similar (in the 0.20–0.22 range) for monolith samples of different thicknesses.

Moreover, researchers found that instead of conventional materials silica aerogels have become more popular due to the highly porous structure, an increase in porosity of polymer structure with the addition of silica aerogel extends the path of sound waves through the material. So, viscous stresses resulting from shearing and friction of air in the material increase that provides to increase sound absorption.

Koh et al. [22] present a simple, practical, and scalable approach to fabricating recycled polyethylene terephthalate (rPET) aerogels for thermal and acoustic insulation applications. The rPET aerogels were successfully developed from recycled PET fibers and polyvinyl alcohol (PVA) and glutaraldehyde (GA) cross-linkers using a freeze-drying process. The effects of various PET fiber concentrations (0.5, 1.0, and 2.0 by wt.%), fiber deniers (3D, 7D, and 15D), and fiber lengths (32 mm and 64 mm) on the rPET aerogel structures and multi-properties were comprehensively investigated. The developed rPET aerogels showed a highly porous network structure (98.3–

99.5%), ultra-low densities (0.007–0.026 g cm^{-3}), hydrophobicity with water contact angles of 120.7–149.8°, and high elasticity with low compressive Young's modulus (1.16–2.87 kPa). They exhibited superior thermal insulation capability with low thermal conductivities of 0.035–0.038 W m^{-1} K^{-1}, which are highly competitive with recycled cellulose and silica-cellulose aerogels and better than mineral wool and polystyrene. The Noise Reduction Coefficient obtained was 0.45 with 2.0 wy. % of rPET.

Table 2. Aerogels physical properties

Material	Noise Reduction Coefficient (NRC)	Thermal conductivity W m^{-1} K^{-1}	Density mg cm^{-3}	Porosity	Reference
Rubber aerogel	0.56	0.035-0.049	91	>90%	[19]
Reinforced Nanofibrous Aerogels (RNFA)	0.41	-	10.76	-	[18]
Pineapple aerogels	0.52	0.030-0.034	13-33	~99%	[20]
Monolithic aerogel	0.20-0.22	-	93-96	-	[21]
rPET	0.45	0.035–0.038	7-26	98.3–99.5	[22]

The acoustic insulation of materials is expressed through Noise Reduction coefficient (NCR) according to ASTM C-423-01 (Standard Test Method for Sound Absorption and Sound Adsorption Coefficients by the Reverberation Room Method), which represents the ability to convert the incident sound/pressure energy into the heat energy and other types of energy by vibration, friction and air viscosity [20]. Table 2 shows some physical properties for synthetized aerogels, as can be observed Monolithic aerogel obtained by Merli et al. [21] exhibits a lower Noise Reduction Coefficient than the other materials, this result can be attributed to the concentration precursors, due to higher density. As higher fiber concentration is used, the structure of PF aerogels becomes more packed and their pore size decreases, hence sound is not able to transmit through the pores easily. Consequently, the energy loss of sound waves increases, which leads to an increase in sound absorption coefficient [12, 20]. Also, it should be noted that most of them are competitive compared with a commercial acoustic foam absorber, Basmel, which has an NCR value of 0.40 [22].

Under this line of research and observing the innumerable advantages that aerogels offer in sound insulation, they also are applied with excellent results as thermal insulators, this being their main international market, due to compared with commercially available thermal insulation materials such as mineral wood, glass wool, polystyrene, polyurethane foam, and cellulose, aerogels offer much higher thermal efficiency with the same thickness. Due to typical insulating materials are limited by the fact that the effective pore sizes are in the micron range and thus do not profit from the suppression of the gas phase transport at ambient conditions. In contrast, aerogels are highly porous materials with pore sizes that can be tuned from the micron to the nanometer range. As a consequence, aerogels show the lowest thermal conductivities at ambient conditions, comparing the K-values for conventional products shown in Table 3 (0.020-0.055 W m^{-1} K^{-1}) and Aspen aerogels (Table 1, K-values ranging 0.0145-0.022 W m^{-1} K^{-1}), taking in consideration that aerogels are a remarkable class of materials with a porosity of more than 90%. In them, the two modes of heat transport, conduction, and convection have been virtually eliminated. Consequently, their thermal conductivity can be reduced to a very low value approaching the air 0.026 W m^{-1} K^{-1} [20].

Even though a wide variety of aerogels with excellent thermal properties can be found on the market, researchers are continuously engaged in finding low-cost and thinner alternatives for insulation in the building and construction industry, because an effective way towards saving energy is to improve the thermal insulation of buildings, especially in hot climates where the energy demand for cooling by air conditioning is comparatively higher. In addition to the need for energy saving, high insulating materials are further justified by improved comfort levels and increased building, furthermore, life thinner insulation materials offer benefits like additional floor area and increasing loading capacity [14, 23].

Table 3. Commonly available thermal insulation materials

Material	Thermal conductivity W m^{-1} k^{-1}	Flammable	Remark
Mineral wool	0.034-0.045	No	Does not melt or support combustion
Glass wool	0.0031-0.043	No	Does not absorb water
Expanded Polystyrene (EPS)	0.029-0.055	Yes	Is difficult to use around imperfections; can become costly
Extruded Polystyrene (XPS)	0.02-0.048	Yes	-
Polyurethane foam	0.020-0.029	Yes	Makes a great sound insulator

To understand the complex heat transportation process in matrix-like aerogels, scientists study the mechanism that takes place, with the purpose of modified conveniently synthesis methods and precursors to improve their thermal properties.

Heat Transfer Mechanisms

The energy (heat) transfer in an insulating material involves three mechanisms: solid conductivity (λ_s), gaseous conductivity (λ_g), radiation transmission in the infrared region (λ_r) and coupling thermal conductivity (λ_c). The sum of these three components gives the total thermal conductivity of the material [14, 24].

$$\lambda_{total} = \lambda_s + \lambda_g + \lambda_r + \lambda_c \quad (1)$$

Solid Thermal Conductivity (λ_s)

Solid thermal conductivity is an intrinsic property of a specific material and refers to heat transferring through the skeletal structure of aerogel consisting of phonon vibration and electron interaction. Typically, the photon heat conduction is mainly related to the density while the electron heat conduction is mainly related to the conductivity of the materials. Solid conductivity and the gas conductivity were proportional to the density as shown below [14, 24]:

$$\lambda_s \propto \rho^{\alpha} \quad (2)$$

Where α is thermal diffusivity related to the joint area between the gel particles in the aerogel determined by experiments and ρ (kg m^{-3}) is the density. As can be seen from equation 2, reducing the density of aerogel can effectively weaken solid heat transferring [24, 25]. For Fricke [26] thermal diffusivity in solid thermal conductivity acquired the value of 1.5.

Gaseous Thermal Conductivity (λ_g)

Gaseous Thermal conductivity is the heat transferring by gas molecule collision. The thermal conductivity as a function of gas pressure is a property that is closely related to the open porosity and the pore size of a porous solid.

In particular, thermal transport via the gas phase significantly changes with gas pressure when the pore size is on the order of the mean free path of the gas molecules. The gas pressure dependence of the gaseous thermal conductivity λ_g is given by the Knudsen-formula [27]:

$$\lambda_g = \frac{\phi\lambda_{g,0}}{1+2\beta K_n} = \frac{\phi\lambda_{g,0}}{(1+2\beta l_{gas}(p_0)p_0/pD)} = \frac{\phi\lambda_{g,0}}{1+p_{1/2}/p} \quad (3)$$

where $\lambda_{g,0}$ is the conductivity of the free gas, β is a constant with a value of about 1.5 that depends on the gas and the accommodation coefficient, and ϕ is the porosity of the sample. K_n is the Knudsen number defined as the ratio of the mean free path l of the atoms or molecules in the gas at a given pressure and the effective pore diameter D; p_0 is a reference gas for which l is known, and $p_{1/2}$ is the pressure at which the value for the gaseous thermal conductivity is half of the maximum value for the free gas ($\phi\lambda_{g,0}$) [27].

A simple expression for the gaseous thermal conductivity was given by Fricke et al. [26] that relates λ_g with a density as follow is expressed:

$$\lambda_g \propto \rho^\alpha \quad (4)$$

Where α is thermal diffusivity in gases that acquired the value of -0.6.

Radiation Thermal Conductivity (λ_r)

The nature of radiative heat transfer within an aerogel, whether it is diffusive or non-diffusive, depends on the optical thickness of the specific type of aerogel. For optically thick aerogels, radiative heat transfer is described by the diffusion of photons. Radiation thermal conductivity (λ_r) means heat transferring by photon transport, which is related to specific extinction coefficient (e), density (ρ), and temperature (T).

$$\lambda_r = \frac{16\sigma n^2 T^3}{3\rho e} \quad (5)$$

where n is the effective refractive index of the aerogel, and σ=5.67×10^{-8} W·m^{-2}·K^{-4} is the Stefan–Boltzmann constant. Because λ_r is proportional to the triple order of T, the increase in temperature will greatly improve the thermal conductivity of aerogels. The results show that when the temperature is higher than 500 °C, the radiation heat conduction plays a major role in the overall heat transfer, which is much higher than the gas phase and solid phase heat transfers and e is the specific extinction coefficient is a material property, and it depends on the chemical composition and structure of the medium. The specific extinction coefficient represents the infrared radiation shielding ability [24].

Coupling Thermal Conductivity (λ_c)

The total thermal conductivity of aerogels can be described by a simple superposition of the different individual heat transport contributions. However, recent investigations reveal that thermal coupling effects can result in a gas pressure-dependent contribution that may be up to three times higher than expected from just a gas phase thermal conductivity, which is predicted by the Knudsen equation at given porosity and pore size [28].

The differences between the thermal conductivity of solid-state and gas are greater; thus, the coupled heat transfer becomes greater correspondingly. Swimm et al. [28] found that the higher the ratio of the intrinsic thermal conductivity of the solid phase to the gaseous thermal conductivity, the stronger the coupling effect. The larger the difference between the two thermal conductivities, the higher is the impact of a thermal shortcut at small gaps in the backbone; thus, the coupling effect increases. The coupling thermal conductivity (λ_c) is caused by the difference between the gas, solid, and radiation thermal conductivities and therefore cannot be controlled directly [24].

Thermal Insulators

Silica Aerogels

Silica aerogels are materials with poor chemical reactivity, very large surface areas (of the order of 1,000 m^2 g^{-1}), unusual porous volumes (greater than 95%), morphologies of monoliths or powders, optical properties (transparent, opaque, or translucent), and very low thermal conductivity in the order of 0.02W m^{-1} K^{-1} at ambient pressure in air and in the order of 0.01 W m^{-1} K^{-1}

when evacuated, they find high added-value applications in the physics of high-energy particles as Cherenkov emitters, transparent and super-insulating double windows, life and space science as well [29, 30].

Tables 1 and 2, summarize the thermal conductivity values of common thermally insulating materials and commercially available aerogels. The range of thermal conductivity is reported for common thermal insulating materials such as polyurethane foam, mineral wool, polystyrene, and glass wool are the highest commercially available silica aerogel materials. The thermal conductivity values reported in Table 2 are significantly lower than air which can be further explained with the Knudsen effect. According to the Knudsen effect, the nanometric pore size of aerogel, which is comparable or lower than the mean free path of air, hinders the thermal conductivity due to the restriction of movement of the trapped gas inside the pore of the aerogel [31].

For silica aerogels, the thermal conductivity decreases with a decrease in density because heat transfer through conduction will be decreased due to a decrease in solid volume in aerogel. Conversely, porosity will show the opposite. Here, the thermal conductivity decreases with an increase in porosity. This happens due to the increase in trapped air as in general thermal conductivity of air is much lesser as compared to any solid material [31].

Hence, to modify the porosity and density of silica aerogel to achieve desired thermal conductivity, various attempts are made by several researchers. The tailoring of thermal conductivity property can be achieved by using different types of precursors or through the synthesis of composites. Thermal conductivity values reported in the literature to be in the range of 0.0221–0.19 W m^{-1} K^{-1} and thermal stability of composite aerogel are documented in the range 280-470 °C. In silica, composites are observed that thermal conductivity and thermal stability properties are strongly dependent upon the fiber content fiber alignment and type of opacifier/whisker [31].

The density and porosity of the aerogel composites are reported to be in the range of 0.05–1.35(g cm^{-3}) and 65.3–93.6%, respectively. In general, the density is increased for aerogel composites as compared to bulk silica aerogel while porosity shows the opposite trend. Similar to bulk aerogel, the thermal conductivity of aerogel composite decreases with a decrease in density.

It is reported that pure silica aerogel is highly transparent in the IR region, particularly from 3 to 8 μm which increases the possibility of radiative heat transfer. So, to improve the thermal insulation property of aerogels for high-temperature applications to be doped with opacifiers such as TiO_2, $K_2Ti_6O_{13}$ whisker, carbon, etc. Thus, transparent aerogel can be fabricated in the form of opaque and it can be useful in spacecraft application because heat transfer by radiation is dominated in space rather than other two modes of heat transfer e.g., conduction and convection. Thermal stability is also depending upon the reinforcing materials with the aerogel matrix.

For the prospect of aerogel based thermal insulation and its challenges can be enlisted the following applications:

- The aerogel composite blankets are widely used as thermal insulation for various ground and aerospace applications. The aerogel composite thermal insulator blanket or wrap are commercially available from vendors such as Aspen, Cabot, etc. and they are mainly used in the industrial pipeline for gas and oil.
- Further, in construction and building fields particularly in European countries and North America.
- Recently, aerogel is also reported to be used as thermal insulation in the textile industry [32].
- Apart from the ground thermal insulation application, bulk aerogel and aerogel composite blankets are also used in space applications. For example, aerogel based thermal insulation blankets are successfully implemented in cryogenic propulsion tanks [33, 34].

Carbon Aerogels

The characteristics of high specific surface area, large porosity, low density, and three-dimensional nanoporous network structure of aerogels make them ideal thermal insulation materials and especially to have good application prospects in the field of aeronautics and astronautics. These advantages encouraged researchers to examine the thermal properties of carbon aerogels at high temperatures in detail. Therefore, carbon aerogel (CA) is expected to become a new generation of ultra-high-temperature and high-performance insulation materials. CA is a porous carbon material that has received considerable attention, especially for extinction coefficients that are much higher than other aerogels e.g., 190 $m^2\ kg^{-1}$ for CA and 20 $m^2\ kg^{-1}$ for pure silica aerogels [35, 36]. The first reported carbon aerogel was synthesized by Pekala [37] via high-temperature pyrolysis of organic aerogels, which was synthesized through sol-gel polymerization of resorcinol (R) and formaldehyde (F) in aqueous solution (so-called RF aerogel) obtained RF aerogels are easily produced in densities ranging from 0.035-0.100 g cm^{-3}. The RF aerogels are composed of interconnected beads with diameters of 7- 10 nm and cell sizes less than 100 nm. Typically, a variety of systems of carbon aerogel have been investigated by altering the reactants. Carbon aerogel properties such as controlled porosity, high surface area, and low density make it ideal for many applications, among them as a thermal insulator or flame retardance [7], especially for the highest thermal stability and can maintain its mesoporous structure even after heat treatment up to 2800°C. In addition, carbon aerogel has much higher specific extinction coefficients than other aerogels and small pores sizes (<100 nm) typically exhibit low thermal conductivity due to the Knudsen effect. The Knudsen effect is observed when gas molecules are confined within pores with diameters smaller than the mean free path of the gas molecules (70 nm) at 1 bar, which induces its high-temperature thermal conductivity to be very low. Therefore, carbon aerogel is potential material for high-temperature thermal insulators [24].

The principal performance hurdle for widespread adoption of CA in thermal insulation applications is their poor matrix- and interface-dominated mechanical performance, particularly under longitudinal compression, in-plane shear, and through-thickness loading. These limitations also result in a susceptibility to defects and damage, which has had a profound effect on the cost of certification and ownership of aerogels. Some reinforcement materials that have been used for the synthesis involve organic fibers [36], ceramic fibers, silica, and polymers [38].

Reinforcement of carbon aerogel with carbon fibers has many advantages. Carbon fibers can significantly improve the mechanical strength and flexibility of the carbon aerogel. Moreover, carbon fibers can also prohibit linear shrinkage of carbon aerogel during the pyrolysis procedure. For example, Fu et al. [39] fabricated carbon aerogel through resorcinol and furfural were polymerized in isopropanol using HCl as a catalyst, and carbon fibers were used to structurally reinforce the architecture of low-density carbon aerogel. The compressive strength of the fiber-reinforced carbon aerogel increased twice as much as that of the pristine carbon aerogel.

Furthermore, the mechanical properties of aerogels generally tend to suffer from high porosity. Appropriate properties control, therefore, is essential to practical applications of carbon aerogels.

Despite, CA aerogels being among the most attractive porous carbon materials currently, the real applications are greatly limited by their high cost, complicated preparation processes, and low mechanical properties. However, [40], reports a very facile route to prepare lightweight but mechanically strong carbon aerogel monoliths (CAMs), based in a facile and low-cost route through the ambient pressure drying. The key to the synthesis is employing commercially available linear phenolic resin as precursor and hexamethylenetetramine (HMTA) as a cross-linking reagent in alcohol solutions. The obtained CAMs possess hierarchical structure containing micropores within the individual particle, and macropores in the interstitial voids between them, aerogels obtained, present the following characteristics: low density as low as 0.07 g cm^{-3}, the high mechanical strength of 0.9-5.0 MPa, and low thermal conductivity (0.032-0.069 W m^{-1} K^{-1}). This method avoids the use of supercritical conditions for drying that are considered highly dangerous and costly and guarantees a balance between the strength and low density of carbon aerogels.

Natural Cellulosic Aerogels

Cellulose, as one of the main components of cellulosic bioresources, is well-thought-out as an alternative for petroleum-based polymers, with enthralling properties to run into the outstanding market for biocompatible, biodegradable, and sustainable materials. Being entirely biocompatible and compostable, cellulosic derivatives, for example, cellulose fibers, cellulose hydrogels and aerogels, cellulosic composites, and cellulose-based films, hold an excessive benefit when contrasted with synthesized polymer equivalents [42, 43].

Recently, cellulose and cellulosic materials have acquired extensive attention for the development of aerogels because of their biodegradability, sustainability, and nanosized structure [43]. Duong Hai Minh in 2016 addresses a novel recyclable cellulosic aerogel and won the 2016 TechConnect Innovation Award. He and his coworkers created the world's first green cellulosic aerogels manufactured from paper waste. The exceptional properties of cellulosic aerogels such as true sustainability and renewability, regeneration, low cost, and holding great capabilities for industrialization. These exceptional properties of cellulosic aerogels are promising for various applications, including energy storage devices, [44] sensors, [45, 46], pollutant treatment, [47, 48] thermal insulation materials [49], and so on.

Characteristics such as renewability, recyclability, non-toxicity, sustainability, simple routes for production, and low thermal conductivity, make cellulosic aerogels an available option for their use in thermal insulation and fire-retardant materials.

In this context, [20] focused efforts to synthetized an aerogel based on cellulose taking advantage that they are considered eco-friendly materials, moreover, a highly porous but stable structure of cellulose aerogels can be created with good mechanical strength and high absorption capacity with the use of a suitable adhesive agent. The raw material that they employed was pineapple fiber as cellulose recourse in combination with polyvinyl alcohol (PVA). The evaluated effects of the fiber concentration on the aerogels' morphology at 0.5, 1.0, and 2 wt. %, mechanical strength, thermal stability, heat, and acoustic insulation were investigated. The excellent heat insulation performance of pineapple fiber aerogels was demonstrated with a thermal jacket for military water. Ice slurry at -3°C and hot water at 90°C were used for heat insulation tests, and they found that the ice slurry temperature can be maintained below 0°C for 6 hours, and the hot water can be maintained above 40°C for 2.5 hours. The thermal insulation abilities were evaluated by K-values, the thermal conductivities for all the developed aerogels were extremely low (0.03-0.034 W m^{-1} K^{-1}) as shown in Table 4 closed to air thermal conductivity of 0.026 W m^{-1} K^{-1} at ambient temperature and pressure, also they identify that k-values increases with pineapple fiber concentration due to a decrease pore volume. The thermal conductivities of pineapple aerogels obtained were comparable to that of good insulation materials such as wool (0.03-0.04 W m^{-1} K^{-1}) and silica aerogels (0.026 W m^{-1} K^{-1}).

Table 4. Summary properties of the pineapple fiber aerogels

Pineapple fiber concentration wt. %	Density mg cm^{-3}	Porosity %	Surface area m^2 g^{-1}	Pore volume cm^3 g^{-1}	Thermal conductivity W m^{-1} K^{-1}	NRC
0.5	12.72 ± 0.38	98.85 ± 0.03	5.556	0.027	0.030 ± 0.001	0.28
1.0	21.02 ± 1.75	98.07 ± 0.16	7.32	0.018	0.031 ± 0.001	0.33
1.5	32.63 ± 2.75	96.98 ± 0.25	3.65	0.012	0.034 ± 0.001	0.52

Moreover, the sound absorption coefficient of pineapple aerogels was investigated and they found that this increased with the fiber concentration, due to as higher fiber concentration was used the structure of aerogel become more packed and their pore size decreases, hence sound is not able to transmit through the pores easily. Consequently, the energy loss of sound waves increases, which leads to an increase in the sound absorption coefficient. The acoustic absorption of all samples reaches a peak at around 2500 Hz before decreasing at higher frequencies. It is noticeable that the 2.0 wt.% pineapple fibers aerogels show the best acoustic absorption. Additionally, it shows a noise reduction coefficient (NRC) of 0.52 with 2 wt.% fiber concentration, indicating the competitiveness of the aerogel to the commercially available product [20].

Nevertheless, there are quite specific deficiencies for the effective utilization of cellulosic aerogels as a result of their internal scarcities such as moisture sensitivity, change of density, and shrinkage due to drying technique. Besides, cellulose fiber is a highly combustible biopolymer and extremely hard to extinguish due to its large specific surface area and high porosity. Compared with other cellulose-based insulating materials, they usually emit an enormous amount of undesired asphyxiated gases during the flammability test with a low limiting oxygen index (LOI) value of nearly 18~19% which proposes critical public safety concerns. Past reports suggest that the incorporation of different inorganic particles enhances the flame retardant properties of cellulosic aerogels. For instance, Donius et al. [50] utilized sodium-montmorillonite (MMT) inorganic particles to form cellulosic aerogel composite finding that the shape retention and flame retardancy functions of the aerogel were remarkably improved up to 800 °C. The latter, because the presence of MMT platelets might reinforce and create a protective thermal barrier slowing down gas diffusion and inhibiting shrinkage of the material.

Disadvantages of Aerogels

Despite the lower thermal conductivities of aerogels compared with some commercially insulating materials (polyurethane foam, chlorofluorocarbons, expanded and extruded polystyrene), their large scale production of aerogels panels is restricted to conventional windows that cannot be equipped with the commonly prepared granulate which is not transparent due to the roughness of its surface. For aerogels synthesis, only frosted glass windows (for example, bathroom, staircase, or ceiling windows) can be used. In these materials, the scattering of the sunlight in the granular layer is rather advantageous, because diffuse lighting of the rooms is achieved [2].

Widespread uses of aerogel materials are restricted at present mainly due to their, high production costs, poor mechanical properties, and Health issues. Supercritical drying is the most expensive and risky aspect of aerogel making process. A highly desirable goal in aerogel preparation is the elimination of the supercritical drying process [14, 32].

Aerogel is a mechanical irritant to the eyes, skin, respiratory tract, and digestive system. Small aerogel particles can potentially cause silicosis, and so forth when inhaled and can induce dryness of the skin, eyes, and mucous membranes. Therefore, protective gear including respiratory protection, gloves, and eye goggles must be worn while handling aerogel. Products are now available, however, which are safe to touch and/or encapsulated to enable them to be worked [32].

Future Perspectives

Analysts forecast the global aerogel market to grow at a CAGR of 35.86% over the period 2014-2019. The high number of patents being filed for new uses of aerogel and the development of new aerogel-based products are expected to contribute to market growth in the coming year. However, to improve development opportunities and to take full advantage of aerogels' excellent thermal insulating abilities, further research needs to be conducted in the areas of [14]:

- Synthesis and characterization of various types of aerogels focused on high-performance textiles.
- Manufacturing methods for further process improvement and cost reduction to add to those already making an impact such as JIOS's rapid one-step and ambient drying process.
- New types of aerogel composites with properties desirable in clothing and footwear.
- Development of new methods to treat fabrics with aerogel particles to reduce fragility and sensitivity to moisture.
- Refinement of thermal measurement techniques as the existing measurements are designed mostly for standard ambient conditions and are not suitable for high-performance insulation materials at extreme temperatures.
- Fabrication of new devices for the measurement of thermal properties in fibrous structures.
- Modeling of convective heat transfer phenomena in fabrics treated with aerogels to enable a better understanding of the thermodynamics under extreme temperature conditions.
- Develop or update standards for thermal measurement in aerogel-based insulation materials used for textile applications, because existing standards do not address the necessary parameters under extreme conditions.

Conclusion

Aerogels are materials with unique properties that can be effective for many applications in different fields of science and technology, taking into consideration, that users can be select the precursors in the synthesis method to improve certain features to make specific aerogels for a particular task. However, their real-world applications are greatly limited by their high cost, complicated preparation process, and low mechanical properties. For this reason, scientists are migrating from supercritical extraction of the solvents toward ambient pressure drying for low-cost preparation. On the other hand, some investigations have been undertaken to strengthen gel skeletons by reinforcing with carbon cloth, carbon fibers, ceramic fibers, silica, and cellulose. Unfortunately, these approaches not only increase the bulk density but also lead to the formation of large cracks during the pyrolysis. So far, it is still a very challenging task to prepare low-density carbon aerogels with integrity structure by ambient pressure drying. Unfortunately, the high mechanical properties always conflict with the low density and high porosity of the materials.

For thermal and acoustic applications, aerogels are competitive in comparison with commercial products such as mineral wood, glass wood, expanded polystyrene, extruded polystyrene, and polyurethane foam, showing similar K-values that air thermal conductivity (0.026 W m^{-1} K^{-1}), also was possible to establish a directly proportional relationship between density, noise coefficient reduction and thermal conductivity. At higher density, higher NCR, and thermal conductivity. The increasing values of thermal conductivity with high density are due to a higher concentration of the solid phase, reducing pore volume, and increasing the capability of the material for heat transport.

References

[1] Ziegler, C., A. Wolf, W. Liu, A.C. Herrmann, N. Gaponic, A. Eychmüller. *Angew. Chem. Int.* Ed. 56 (2017) 13200 – 13221.

[2] Hüsing, N., U. Schubert. *Ullmann's Encyclopedia of Industrial Chemistry.* 1 (2012) 621-646.

[3] Leventis, N., A. Sadekar, N. Chandrasekaran, C. Sotiriou-Leventis. *Chem. Mater.* 22 (2010) 2790–2803.

[4] Vareda, J. P., A. Lamy-Mendes, L. Durães. *Micropor Mesopor Mat.* 258 (2018) 211-2016.

[5] Gurav, J. L., I. K. Jung, H. H. Park, E. S. Kang, D. Y. Nadargi. *J. Nanomater.* (2010) 1-11.

[6] Maleki, H., L. Durães, C. A. García-González C., P del Gaudio, A. Portugal, M. Mahmoudi. *Adv. Colloid Interface Sci.* 236 (2016) 1-27.

[7] Lee, J. H., S. J. *Park. Carbon.* 163 (2020) 1-18.

[8] Li, H., Z. Wang, X. Liu, F. Cui, C. Chen, Z. Zhang, J. Li, L. Song, R. Bai. *Chem. Phys. Lett.* 755 (2020) 137805.

[9] Maleki, H., *Chem. Eng. J.*300 (2016) 98-118.

[10] Batista, M. P., V. S. S. Gonçalves, F. B. Gaspar, I. D. Nogueira, A. A. Matias, P. Gurikov. *Int. J. Biol. Macromol.* 156 (2020) 773-782.

[11] da Silva, F. T., J. P. de Oliveira, L. M. Fonseca, G. P. Bruni, E. da Rosa-Zavareze, A. R. Guerra-Dias. *Int. J. Biol. Macromol.* 155 (2020) 6-13.

[12] Yanga, W., J. Liua, Y. Wanga, S. Gaoa. *Energ Building.* 206 (2020) 109583.

[13] Koebel, M., A. Rigacci, P. Achard. *J Sol-Gel Sci Technol.* 63 (2012) 315-339.

[14] Thapliyal, P. C., K. Singh. *J. Mater.* (2014) 1-10.

[15] Company website: https://www.cabotcorp.com/company (11/10/2020).

[16] Company website: https://www.aerogel.com (11/10/2020).

[17] Company website: http://www.nanopore.com/company.html (11/10/2020).

[18] Cao, L., Y. Si, Y. Wu, X. Wang, J. Yub, B. Ding. *Nanoscale.* (2019).

[19] Ba Thai, Q., R. O. Chong, P. T. T. Nguyen, D. K. Le, P. K. Led, N. Phan-Thiena, H. M. Duong. *J. Environ. Chem. Eng.* 8 (2020) 104279.

[20] Do, N. H. N., T. P. Luu, Q. B. Thai, D. K. Le, N. Do Quyen Chau, S. T. Nguyen, P. K. Le, N. Phan-Thien, H. M. Duong. *Mater. Chem. Phys.* 242 (2020) 122267.

[21] Merli, F., A. M. Anderson, M. K. Carroll, C. Buratti. *Appl. Acoust.* 142 (2018) 123-131.

[22] Koh, H. W., D. K. Le, G. N. Ng 1, X. Zhang, N. Phan-Thien, U. Kureemun, H. M. Duong. *Gels.* 4 (2018) 1-13.

[23] Grand View Research. Aerogel Market Size, Share & Trends Analysis Report By Form (Blanket, Particle, Panel, Monolith), By Product (Silica, Carbon, Polymers), By End Use, By Technology, By Region, And Segment Forecasts, 2018-2025. 2018 September 26, 2020]; Available from: https://www.grandviewresearch.com/ industry-analysis/aerogelmarket.

[24] Hu, L., R. He, H. Lei, D. Fang. *Int. J. Thermophys.* 40:39 (2018) 1-25.

[25] Bock, V., O. Nilsson, J. Blumm, J. Fricke. *Non Cryst Solids.* 185 (1995) 233-239.

[26] Fricke, J., *Proceedings in Physics.* 6 (1986) 94–103.

[27] Reichenauer, G., U. Heinemann, H. P. Ebert. *Colloid Surface A.* 300 (2007) 204-210.

[28] Swimm, K., G. Reichenauer, S. Vidi, H. P. Ebert. *J Sol-Gel Sci Technol.* (2017) 1-9.

[29] Pajonk, G. M., *Colloid Polym Sci.* 281 (2003) 637-651.

[30] Soleimani Dorcheh, A., M. H. Abbasi. *J. Mater. Process*. 199 (2008) 10-26.

[31] Hasan, M. A., R. Sangashetty, A. C. Mary Esther, S. B. Pati, B. N. Sherikar, Arjun Dey. *J. Inst. Eng. India Ser. D.* (2017) 208.

[32] Venkataraman, M., R. Mishra, T. M. Kotresh, J. Militky, H. Jamshaid. *Text. Prog.* 48:2 (2016) 55-118.

[33] Burchell, M. J., G. Graham, A. Kearsley, Cosmic dust collection in aerogel. *Annu. Rev. Earth Planet Sci.* 34, 385–418 (2006).

[34] Fesmire, J. E., Aerogel insulation systems for space launch applications. *Cryogenics* 46, 111–117 (2006).

[35] Wiener, M., G. Reichenauer, F. Hemberger, and H. P. Ebert. *Int J Thermophys.* 27:6 (2006) 1826-1843.

[36] Feng, J., C. Zhang, J. Feng, N. Zhao, ACS. *Appl. Mater. Interfaces.* 3 (2012) 4796.

[37] Pekala, R. W., *J. Mater. Sci.* 24 (1989) 3221-3227.

[38] Nguyen, S., D. B. Anthony, H. Qiana, C. Yueb, A. Singha, A. Bismarck, Milo S. P. Shafferc, E. S. Greenhalgh. *Compos Sci Technol.* 182 (2019) 107720.

[39] Fu, R., B. Zheng, J. Liu, *Adv. Funct. Mater.* 13 (2010) 558.

[40] Jia, X., B. Dai, Z. Zhu, J. Wang, W. Qiao, D. Long, L. Ling. *Carbon*. 108 (2016) 551-560.

[41] Wang, S., A. Lu and L. Zhang. *Progress in Polymer Science* 53 (2016) 169-206.

[42] Zaman, A., F. Huang, M. Jiang, W. Wei, Z. Zhou. *Energy and Built Environment.* 1:1 (2020) 60-76.

[43] Tan, C. B., B. M. Fung, J. K. Newman, C. Vu. *Organic Aerogels with Very High Impact Strength.* (2001).

[44] Li, H., L. Song, C. Sun, R. Li, Y. Fu, H. Zhang, A. Yang, H. Liu. *Sci Eng Compos Mater.* 25:6 (2018) 1107-1114.

[45] Qin, J., L. Chen, C. Zhao, Q. Lin and S. Chen. *J. Mater. Sci.* 52:14 (2017) 8455-8464.

[46] Zhong, Y., Y. Kong, X. Shen, S. Cui, X. Yi, J. Zhang. *Micropor Mesopor Mat.* 172 (2013) 182-189.

[47] Jiang, F., Y. L. Hsieh. *J. Mater. Chem.* 2:18 (2014) 6337-6342.

[48] Rafieian, F., M. Hosseini, M. Jonoobi, Q. L. Yu. *Cellulose*. 25:8 (2018) 4695-4710.

[49] Guo, W. W., X. Wang, P. Zhang, J. J. Liu, L. Song, Y. Hu. *Carbohydr. Polym.* 195 (2018) 71-78.

[50] Donius, A. E., A. Liu, L. A. Berglund, U. G. K. Wegst. *J Mech Behav Biomed Mater.* 7 (2014) 88-99.

Chapter 5

Synthesis of Silica Aerogels

Misael García Ruiz*
Universidad Autónoma del Estado de México, Centro Conjunto de Investigación en Química Sustentable UAEM-UNAM

Abstract

Thanks to their excellent characteristics including high surface area, low density, open pore structure, and excellent insulation properties, the aerogels have been intensively investigated since the past decades for a diverse range of applications. Aerogels are usually synthesized by sol–gel method, since this is a simple and widely studied technique. However, there are other synthesis techniques that were little described in the previous literature. On these grounds, this chapter aims to describe the synthesis techniques of aerogels and to offer some future perspectives for their preparation, modifications and applications.

Keywords: silica aerogels, sol-gel process, synthesis, gelation, self assembly, drying

Introduction

Silica aerogels were first prepared in 1931 [1]. These materials present a cross-linked network structure of ultra-low density. Such solid-state materials are generally synthesized by sol-gel method at low temperature [2, 3]. Silica aerogels are materials with unusual properties such as high specific surface

* Corresponding Author's Email: misagr89@gmail.com.

In: Aerogels: Properties and Applications
Editor: Alina Iuliana Pruna
ISBN: 978-1-68507-788-4

area (500–1200 m^2/g), high porosity (80–99.8%), low density (~0.003 g/cm^3), high thermal insulation value (0.005W/mK), ultra-low dielectric constant (k = 1.0–2.0) and low index of refraction (~1.05) [4-7]. It was reported that a native silica aerogel with an average density of about 120 mg cm^{-3} can be completely crushed by applying a stress of about 31 kPa. In addition, aerogels show hydrophobicity, low thermal conductivity, and optical transparency.

The fragile nature of silica aerogels became beneficial in several space missions, e.g., in capturing hyper-velocity particles of cosmic dust from the tail of a comet. Transparent aerogels have been prepared as hyper-velocity particle capture in space (cosmic particles) [8]. Due to their such unusual characteristics, much attention has been given to silica aerogels in recent years for their use in several technological applications including Cherenkov radiators in particle physics experiments [9, 10]. Aerogels had little practical value until 1974, when Cantin et al. [10] reported on the use of silica aerogels as detectors in Cherenkov radiation.

Aerogels have been used or considered for use in laser experiments, waste management, for molten metals, for optics and light guides, electronic devices, capacitors, imaging devices and pesticides. Initially, silica aerogels were used as thermal insulators thanks to their exceptionally low thermal conductivity [11, 12]. SiO_2 aerogels were employed as catalyst support for polymer electrolyte membrane fuel cells. Recently, an increasing interest in aerogels could be observed in pharmaceutical sciences [13, 14], food related technologies [15], biosensors [16] and biotechnology [17]. Other aerogels have been demonstrated as alternatives in battery electrodes [18], catalyst supports [19], oxygen and humidity sensors [20], and as adsorbents for environmental clean-up [21] thanks to their large internal surface areas and facile tailoring of their surface chemistry. Slowing et al. [22] reported that silica aerogels were more biocompatible as compared to many other materials for drug delivery systems. Some studies have shown that the use of the silica aerogels has important effects on the absorption, the drug carrier, and the release rate. On the other hand, the high surface area of aerogels gives them a great potential to be employed as a catalyst or a catalyst carrier. Aerogels aid in heterogeneous catalysis when the reactants are either in gas or liquid phase [6]. For example, synthesis of nitrile from hydrocarbons using nitric oxide (NO) [23] or conversion of isobutene into methacrylonitrile by reacting it with NO over zinc oxide aerogel [24].

Synthesis of Silica Aerogels by Sol-Gel Method

The synthesis of silica aerogels has received significant attention especially during the last two decades. Aerogels are mainly synthesized by the sol-gel process by using silicon alkoxides as precursors. The synthesis of aerogel involves the formation of primary nanoparticles or the building blocks by hydrolysis which is followed by the condensation of the building blocks to form a network structure and finally, the removal of pore liquid without disturbing the network structure [25]. Kistler (1931) reported on the preparation of silica aerogels by using a sol-gel method combined with supercritical drying [1]. Kistler used water as starting material. After exchanging water in the gel with alcohol, the gel was subjected to alcohol supercritical drying. This process involves the transformation of molecular precursors into highly cross-linked inorganic or organic gels that can then be dried by using special techniques (i.e., supercritical drying, freeze drying, etc.) aimed to preserve the tenuous solid network [26, 27]. Kistler also prepared aerogels of alumina, iron oxide, tin oxide, and so on.

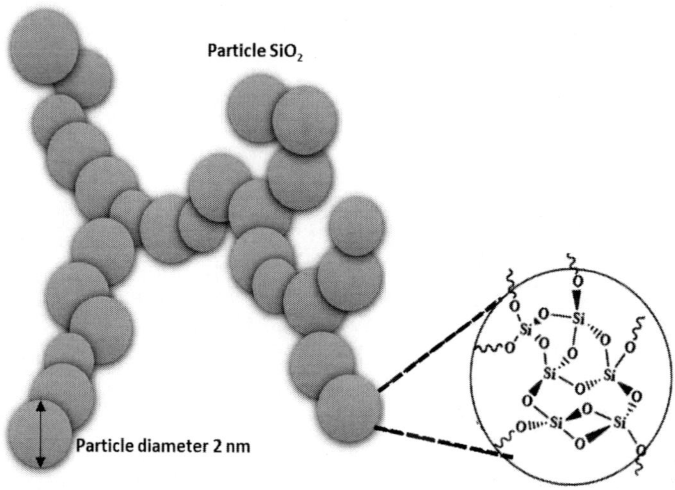

Figure 1. Silica aerogel network.

Sol–gel science dates back over 150 years. The term "sol–gel" was first created by Graham in 1864 during his work on silica sols. Sol–gel precursors are mainly silicon alkoxides which can be obtained in a high degree of purity. It involves the transition of a system from a liquid "sol" into a solid "gel" phase. In a sol, colloidal particle with diameters in the range of 1 - 1000 nm are dispersed in a liquid (Figure 1). A colloid is, by definition, a suspension of small (1-1000 nm) particles that diffuse by Brownian motion. A gel consists of a sponge-like, three-dimensional solid network whose pores are filled with another substance (usually a liquid) [28]. The sol-gel process can ordinarily be divided into the following steps: forming a solution, gelation, aging, drying, and densification (Figure 2). A few of the important advantages of the sol-gel process worth to mention include its simplicity and the fact that it is an economic and effective mean of producing high-quality materials. Sol-gel processing has found application in the production of high-quality glasses for optical components and fibers, thin film coatings, and fine oxide powders [6, 29-31].

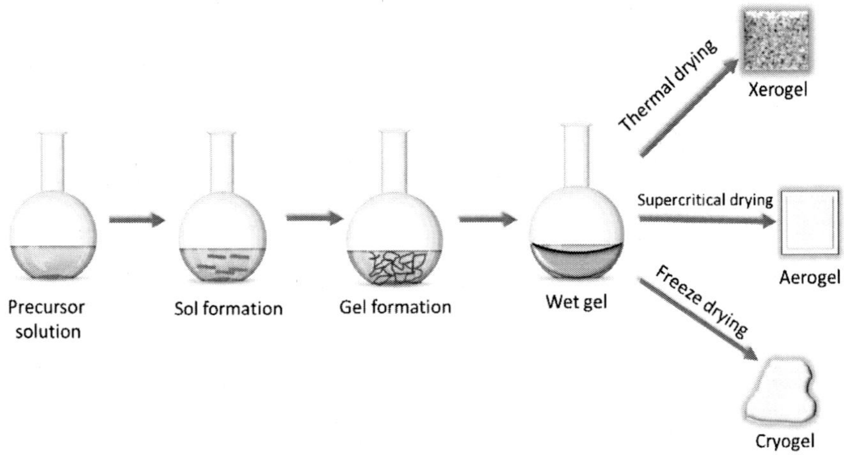

Figure 2. Basic synthesis scheme for aerogel.

Sol-Gel Synthesis Steps

Synthesis of silica aerogels consists of three steps: (1) gel formation or gelation, (2) aging and (3) drying of the gel. Most aerogel synthesis methods follow two main reactions: hydrolysis and condensation, and the formation of a sol and a gel. Sol particles are formed by chemical reactions from dissolved molecular precursors. Their condensation results in the formation of a gel [28]. All methods of aerogel production involve these three general steps. Additional procedures also need to be undertaken to tailor the final product structure. The sol-gel reaction is a versatile process that allows the tailoring of the gel nanostructure by adjusting the reaction key parameters. There are several strategies to influence the reaction kinetics of hydrolysis and condensation reactions including the activity of the metal alkoxide, the water/alkoxide ratio, solution pH, temperature, nature of the solvent, and used additive. Hydrolysis and subsequent condensation reactions are initiated starting from aqueous solutions of ionic precursors by changing the pH or the temperature [29]. Following the addition of stoichiometric amount of water, the hydrolysis reaction is catalyzed by the addition of an acid or a base generating Si-O-Si bonds. Polycondensation reactions which lead to the formation of silica gel start simultaneously with the hydrolysis. Alkoxide based sol–gel synthesis avoids the formation of undesirable salt by-products and allows a much greater degree of control of the final products.

The pH value is a decisive parameter for the relative rates of hydrolysis and condensation of alkoxysilanes for precursors of the type $Si(OR)_4$ in which OR designates alkoxide groups. R is generally CH_3 or C_2H_5. The influence of pH on the structure and properties of aerogels has been studied in detail [32, 33]. The lowest reaction rate for hydrolysis takes place at pH = 7 and for condensation at around 4.5 pH. Under acid-catalyzed conditions (pH < 5), the hydrolysis is favored, while the condensation reactions are limiting [34, 35].

Reactions (1) – (3) describe the chemical reactions taking place during the sol-gel processing of metal alkoxides.

Hydrolysis

$$M - OR + H_2O \rightleftarrows M - OH + R - OH \tag{1}$$

Water condensation

$$M - OH + HO - M \rightleftarrows M - O - M + H_2O \tag{2}$$

Alcohol condensation

$$M - OR + HO - M \rightleftarrows M - O - M + R - OH \tag{3}$$

The synthesis of aerogels begins with the preparation of a mixture of precursors and solvents. When hydrolyzed, these precursors and solvents react with each other and form a colloidal solution of small polymeric particles and/or dissolved polymer clusters [36]. The primary particles grow and agglomerate, forming larger secondary particles that bond with each other and result in the solid three-dimensional (3D) network of the gel. Sol–gel precursors are mainly silicon alkoxides which can be obtained in a high degree of purity. These include tetramethoxysilane (TMOS), tetraethoxysilane (TEOS), polyethoxydisiloxane (PEDS), methyltriethoxysilane (MTES), or propyltrimethoxysilane (EDAS) [37]. Some examples of silicon alkoxides and their structure are shown in Table 1. Metal alkoxides are members of the family of metalorganic compounds, which have an organic ligand attached to a metal or metalloid atom. The alkoxysilanes are insoluble in water, thus a mutual solvent must be normally used as a homogenizing agent. Among the classes of solvents, alcohols are largely used, but other polar (H_2O, acetone) and non-polar solvents (tetrahydrofuran, dioxane, benzene) may be used, as well [33, 34, 38, 39]. The density of the final aerogel can be controlled by varying the ratio of solvent to precursor in the initial solutions. The amount of solvent is the one to define the pore volume within the silica network. Thus, a high solvent to precursor ratio could result in a low-density silica aerogel, i.e., the effect of excess water content [39, 40]. Dorcheh [37] and Wagh et al., [42] investigated the influence of varying precursor material by preparing aerogels from TEOS, TMOS and PEDS. Their work concluded that silica aerogels of different physical properties were obtained by using various precursors and indicated that TMOS generated a gel of higher surface area, with narrow, more uniform pores, and exhibited lower thermal conductivity than those obtained from PEDS and TEOS. In addition, MTES-derived silica aerogels showed flexible properties due to the flexibility of the final glass network and exhibited buoyancy in water because of the hydrophobic property imparted by the methyl group [43].

Table 1. Common Silica Precursors and their chemical structures

Precursors	Chemical structure
Tetramethoxysilane (TMOS)	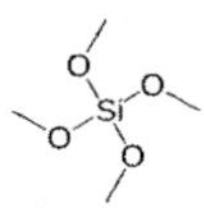
Tetraethoxysilane (TEOS)	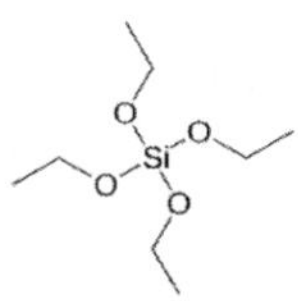
Precursors	Chemical structure

Polyethoxydisiloxane (PEDS)	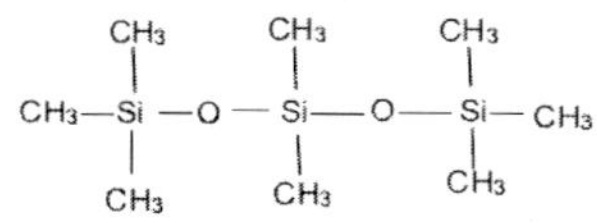
Methyltriethoxysilane (MTES)	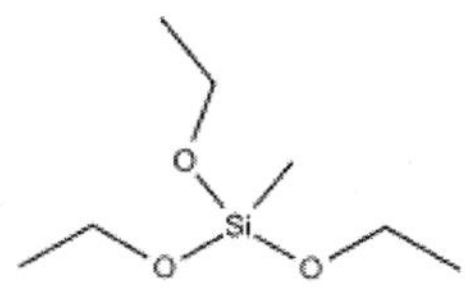

Table 2. Precursors for silica aerogel synthesis

Hydrophilic precursors	Hydrophobic precursors
Tetramethoxysilane (TMOS)	Methyltrimethoxysilane (MTMS)
Tetraethoxysilane (TEOS)	Methyltriethoxysilane (MTES)
Sodium Silicate (Na_2SiO_3)	

Silica aerogels have been studied owing to their hydrophobic nature. When the surface energy of a solid is low, it repels the water from its aerogel surface and vice versa, thus by showing hydrophobicity or hydrophilicity. Currently, hydrophobic surfaces are used in industry for a variety of applications including hydrophobic coatings for rust prevention, oil removal from water, management of oil spills, and chemical separation processes to separate non-polar and polar compounds. As mentioned above, silica aerogels can be either hydrophobic or hydrophilic as depending on the synthesis process. The silanol polar group in the aerogel network (Si-OH) causes the absorption of water and leads to a hydrophilic behavior. Hydrophobic aerogels obtained from the precursor without surface chemical modification are called hydrophobic precursors, and that of hydrophilic are called hydrophilic precursor as provided in Table 2 [6, 44].

In this respect, different types of hydrophobic silica aerogels have been synthetized to adsorb miscible and immiscible organics such as ethanol and toluene [45] and the results showed they absorbed crude oil of about 3.5 times of their weight [46].

Aerogels with hydrophobic and oleophilic surfaces offer the possibility of improving the absorption of pollutants [47]. Hydrophobic silica aerogels contain a high capacity for adsorption of non-soluble organic compounds while hydrophilic silica aerogels appear to be more efficient for the adsorption of soluble organic compounds in water [48]. Nano-cellulose aerogels treated with octyltrichlorosilane by using vapor phase deposition were reported for the fabrication of hydrophobic surfaces for the separation of oil from oil/water mixture [49]. A second approach for increasing the hydrophobicity consists in the modification of the aerogel surface after drying. The surface of hydrophilic aerogels can be modified by reaction with gaseous methanol [50]. Liu et al. [51] found that hydrophobic silica aerogels contain a high adsorption capacity towards poorly soluble organic compounds while hydrophilic silica aerogels were more efficient towards adsorption of soluble organic compounds from water. The adsorption properties of silica aerogels in this study were reported to remaine stable after 5 adsorption/desorption cycles.

The kinetics of the sol–gel reactions is slow at room temperature and often requires several days to reach completion. For this reason, acid or base compounds catalysts are added as catalysts [52]. Acid catalysts can be any protic acid, such as HCl [53], and ammonium fluoride as basic catalysis [34]. Both catalysts lead to a wider distribution of larger pores and a lower thermal conductivity. Acid hydrolysis usually requires longer times (up to 3 days) before the catalyst addition [54, 55]. On the other hand, the base catalysis proceeds faster, and allows the shaping of uniform, branched gels with a large pore volume and greater tensile strength due to high condensation rate. However, base-catalyzed gels display higher densities in comparison to gels prepared with acidic solutions [35].

Gelation

The ageing processrepresents an additional chance of gel network to further grow in the gelation solvent, this process being termed as "ageing." This ageing process strengthens the gel, so that minimum shrinkage occurs during the drying step [56]. However, the aerogels prepared by the sol–gel method are amorphous and require further calcination which leads to the loss of many unique properties of aerogels [57, 58].

Gelation starts to occur when secondary particles link to each other, resulting in neck-regions between them, and a mesoporous network. After gelation, the gel is left undisturbed in the solvent to complete the reaction. After the completion of reaction, the aerogel product is formed.

In fact, the gelation solvent inside the pores contains the reactive species (such as -OH) or unreacted monomers which can mutually condense onto the network. During the ageing process, two mechanisms can take place at different rate: (1) neck growth originated in transportation of silica dissolved from particle surfaces to the neck region between silica particles and (2) accumulation of dissolved smallest particles on the more active network sites related with inequality in surface energy which is known as Ostwald ripening [5]. Small additional silica particles, resulting from the ageing solution, which are merged within silica network dissolve and re-precipitate between larger particles within the chain, reinforcing the gel complex [59]. There is not an activation energy that can be measured, nor can one precisely define the point where the sol changes from a viscous fluid to an elastic gel. The change is thus, gradual, as more and more particles become interconnected. All subsequent stages of processing depend on the initial structure of the wet gel formed during reaction gelation [63]. In this respect, many studies have focused on the process of ageing of aerogels and its effects on the solid framework. Hæreid et al., [60, 61] explored the effects of ageing time, temperature, and pH of wet gels, aged in either TEOS or Ethanol. They observed that TEOS proved to be more favorable ageing solution than ethanol. Moreover, the authors have found that an increase in pH of pore liquid, prior to ageing, results in a shorter gelation period. Finally, the nature and surface tension of solvent employed during the washing of the gel was found to have marked influence on the linear shrinkage of the aerogel [36].

Drying

The final, and most critical process in the synthesis of silica aerogels is the drying step. This process involves the removal of the liquid from the pores of the gel without deforming the solid structure. At the end of the drying process a dry aerogel with a microstructure almost identical to that of the wet gel is obtained. The morphological structure of the final product mainly depends on the choice of pathway applied for the drying of solvents and / or water in the pores from the liquid state to vapor state. Depending on the drying conditions, aerogels, xerogels or cryogels can be obtained. Several drying procedures are used to dry the wet gels, such as supercritical drying (using i.e., alcohol, acetone, or CO_2) [62], ambient pressure drying or evaporation [63] and freeze-drying [64].

If the liquid in a wet "gel" is removed under supercritical conditions, a highly porous and extremely low-density material called "aerogel" is obtained [65]. Drying by evaporation under normal conditions gives rise to capillary pressure that causes shrinkage of the gel network. The resulting dried gel called "xerogel" ("xero" means dry), is often reduced in volume by a factor of 5 to 10 as compared to the original wet gel [66]. Finally, the silica cryogels are synthesized by freeze-drying that includes a sublimation process in pursuit of freezing the liquid in the gel pores.

Supercritical drying is the most common method applied for the drying process of aerogels. The supercritical drying process has been proposed by Kistler (1932) to dry, without textural modification, very tenuous solids wetted with a solvent [67]. In the supercritical fluid drying method, the liquid in the pores is removed above its critical temperature (TCr) and pressure (PCr), i.e., in supercritical state. In this condition, there are no liquid–vapor interfaces and, thus, no capillary pressure gradients arise [5, 68]. As a result, the pore liquid converts to a gaseous phase and thus, it is evaporated. Supercritical drying can be completed at both ambient temperature (low temperature supercritical drying) or at elevated temperatures (high temperature supercritical drying) [37].

As depending on the process temperature, the supercritical fluid drying processes are divided into HOT and COLD ones. In the HOT process, the organic solvents (generally referred to the synthesis alcohols) are in their supercritical state, i.e., near 260 °C if ethanol or methanol are used while in the COLD process, the synthesis solvent is extracted with supercritical CO_2 at a temperature slightly above the critical temperature of CO_2 (i.e., ~31°C) [69];. In the COLD process, the liquid part of the wet gel must be exchanged with CO_2, either in the liquid state [70], or directly in the supercritical state. Therefore, by the removal of pore liquid under supercritical condition, the porous network structure of the wet gel can be preserved without experiencing a considerable shrinkage. However, the supercritical drying usually involves high temperature and pressure as depending on the pore liquid. The most attractive solvent is the carbon dioxide. CO_2 is chemically unreactive, and its critical temperature is close to room temperature. Using CO_2 allows the gels which do not suffer temperatures higher than 100 °C to dry under suitable safety conditions [36]. In comparison to other drying methods, the supercritical drying is a more effective one for preventing the shrinkage or the collapse of the mesopores. However, one main disadvantage for the supercritical CO_2 drying is that it is a time-consuming procedure [71].

The evaporation process results in a material known as a xerogel. Throughout this practice, the liquid is evaporated from the gel without additional treatment, causing capillary pressure to increase due to high surface tension, which can result in fractured gels. In principle, this process involves a series of solvent exchange processes and the chemical modification of the inner surface of the aerogel with nonpolar groups. Such a surface modification with the nonpolar groups on the pore walls prevents the new condensation reactions during the pore contraction upon the capillary stresses [36, 72]. Consequently, the xerogels experience a great deal of shrinkage during the drying process and they exhibit a low porosity [38]. In recent years, the ambient pressure drying attracted increased interest as its production costs are very low because it doesn't require dangerous high-pressure equipment; however, several steps of solvent exchange and chemical reaction are deemed necessary. For example, Rao et al. prepared a silica aerogel with improved properties in terms of low density, high hydrophobicity, low thermal conductivity, high porosity, and high optical transmission by a two-step drying process at ambient pressure from TEOS precursor [73].

The commercial products are typically produced following a supercritical fluid drying process, i.e., recently aerogel-enhanced blankets have been produced at ambient pressure [74, 75]. In the case of drying at ambient pressure, an effect called "spring back" intervenes, in which the gel contracts by an amount equal to the volume of the liquid that evaporates from the pores, due to the development of capillary forces, when the air-displaced liquid diffuses out of the pores of the gel [38].

Finally, the freeze-drying is another option for the drying process of aerogels. Freeze-drying consists of lowering the temperature to induce the liquid solidification (crystallization) [76]. One of the challenging points of this method is the long aging period to be able to stabilize the gel network. This technique removes excess solvent by sublimation. However, due to crystallization of the solvent within the pores, this process usually leads to cracked or even powder-like silica products with very large pores [54]. The porous material obtained by this method is a cryogel and their surface area, and pore volume will be slightly smaller than aerogels. In fact, cryogels obtained from the same gel have 20% less porosity and almost half the specific surface area. Cryogels are, therefore, only obtained as powders [36]. Unfortunately, freeze-drying has some disadvantages including the fact that the aging period must be prolonged so as to stabilize the network which may be further destroyed by crystallization of the solvent in the pores [38].

Other techniques such as microwave drying, and vacuum drying have been applied to obtain aerogels with high surface area and porosity [71]. Aerogels obtained by these procedures show comparable structures, but with smaller interconnected macropores as compared to those prepared by freeze-drying. Microwave drying is a time-saving method to prepare mesoporous and 3D interconnected macroporous aerogels [77].

Surface Modification of Aerogels

Once the aerogel is obtained, further modifications are possible, either by changing the chemistry of the aerogel surface or by changing the composition of the entire aerogel. These changes are generally based on the modification of the surface of the aerogel by the infiltration of molecules with functional groups or doping elements by means of a gas phase [36]. Materials prepared from wet gels generally have external and internal surfaces covered by final hydrolyzed groups (M-OH). These hydroxyl groups on the surface can be changed with other functional groups using hydrolyzable molecular precursors containing the desired functionality [78].

The resultant surface-modified aerogels are applicable in a wide range of application fields including adsorption catalysis, microelectronics, optoelectronics, thin film technology, protective coatings, biosensors of chemical sensors, nanomaterials, and bioactive surfaces, in addition to cell adhesion and adsorption of protein adsorption/drug release, among others [79, 80]. The inclusion of catalysts or nanoparticles, i.e., magnetic of TiO_2, in nanostructure of gel to render additional properties or a magnetic or photocatalytic functionality to the pristine aerogel has been considered, as well [81]. The inclusion of particles was found to affect the aggregation process and therefore, it modifies the porous structure. Thus, it is possible to adjust the density, the mechanical properties, the pore size distribution, and the permeability of the composite aerogels by adding silica powder [66]. After aerogel synthesis, an additional step may be applied to increase the bulk density either by sintering at a high temperature (1000 °C) [82] or by isostatic pressure [83].

In addition to previously mentioned modifications, a supplementary step such as carbonization is sometimes applied, as it is the case of carbon aerogels prepared from polymeric aerogels. In the carbonization stage, the organic aerogel is transformed into an aerogel composed of carbon. For this aim, the temperature is increased above the pyrolysis temperature of the organic compound that forms the gel. The amount of heat to be applied controls the degree of carbonization and the content of residual elements. Carbonization is often an exothermic process. The carbonization process was firstly introduced by Pekala [85] in his pioneering works after the development of resorcinol-formaldehyde (RF) aerogels in 1989. It has continuous porosity, an ultrafine cell/pore size (<50 nm), high surface areas (400–900 m^2 g^{-1}), and a solid matrix composed of interconnected colloidal-like particles or fibrous chains with characteristic diameters of 10 nm [85, 86].

Carbon is deposited as graphite on the inner surface of the aerogels, and even nanometer-sized carbon fibers, rings, and tubes were found. Carbon aerogels can be activated after their preparation [87] and modified on their surface or doped with metals [88]. Depending on the metal dopant and the pyrolysis temperature, carbothermal process can give rise to metal-doped monoliths with nanostructured composition or carbides. Pyrolysis reactions are most important for preparing carbon aerogels, which are of particular interest because they are the first electrically conducting aerogels [28].

On the other hand, the synthesis of non-siliceous aerogels by the sol-gel method has been reported by several authors. In 1975, Yoldas [89] reported for first time Al_2O_3 aerogels from alkoxides of alumina. Al_2O_3 aerogels are interesting for applications as catalysts and catalyst supports, and they are also used for the removal of mustard gas. Synthesis of titania aerogels have been achieved from the corresponding ethoxide, isopropoxide or butoxide by classical sol–gel approaches in aqueous alcohol or ketone mixtures. Upon the calcination, the TiO_2 aerogels are mainly used for applications in photocatalysis, solar cells and in monolithic form as columns for chromatography [90]. Vanadium oxide is a promising cathode material in lithium-ion batteries due to its capability to intercalate Li ions. For such an application, the porous structure of a vanadium aerogel should be highly beneficial. Several groups investigated the electrochemical properties of V_2O_5 aerogels for energy storage [91].

Alternative Synthesis Methods for Silica Aerogels

Self-assembly is a potential tool for the preparation novel modified aerogels with graphene or metals. This method is based on building large and complex aerogels systems at a macroscopic level and complex structures can be obtained [23, 92]. Self-assembly has been recognized for a long time as one of the most effective strategies for “bottom-up” nanotechnology. With unique structure and properties, graphene naturally becomes a versatile nanoscale building block for self-assembly to achieve novel structures and functionalities [93]. Recently, 3D graphene aerogels have been fabricated through self-assembly and thus, they expanded the applications of graphene [94]. Moreover, this method appears advantageous over sol–gel one because of the high crystallinity of the final aerogel. Arachchige et al. [95] reported for the first time the nanoshell based metallic aerogels obtained by self-assembly of a series of bimetallic nanoshells into monolithic gel structures.

On the other hand, a variety of techniques are available for the production of round gel (micro) particles, i.e., emulsion-based processes, as recently reported by Ganesan et al. [96]. In their work, spherical aerogels as well as other forms could be obtained in a millimeter size (monolithic aerogels).

The emulsion process consists of four basic steps: (1) the preparation of the disperse phase by the sol–gel process; (2) the emulsification of the disperse phase in a continuous phase (oil, immiscible with the first one); (3) the cross linking/gelation reaction within the dispersed phase (liquid microdroplets) to form stable gel microspheres; (4) the CO_2 supercritical extraction of the oil–gel dispersion to obtain the final microsphere aerogel particles. [97, 98]. This method appears advantageous for synthesizing aerogels in micro-spherical form by maintaining the internal structural properties identical to that of monolithic aerogels [22]. This synthesis technique has been widely studied by Alnaief et al. [98]. They synthesized micro-spherical silica aerogel particles by combining the sol–gel process with the emulsion formation followed by supercritical extraction of the oil-particles dispersion. However, the most common technique used for the preparation of aerogel nanostructure is still represented by the sol–gel method as this is a simple and widely studied.

Future Perspectives

The chemistry of aerogels is very flexible thanks to their pore size, high surface area (>150 m^2/g) and open porosity (typically 95–99.99%); furthermore, different functional groups can be implemented in order to provide with desired mechanical and physico-chemical properties. Thus, aerogels have attracted increasing interest from both the academic and industrial communities in the past two decades. Aerogels might be used for a range of applications in a wide spectrum of areas, ranging from construction, chemical engineering, thermal insulation, catalysis, physics, environmental remediation, optical devices, and hypervelocity particle capture and more recently in life sciences and medicine [99-106].

The physical and chemical modification of an aerogel surface to improve the adsorption process showed a great potential for the future. It is a clear fact that aerogels have extraordinary characteristic features like very low thermal conductivity, translucent structure, good sound insulation etc. However, aerogels could be extended to any kind of highly porous solid materials with low density, large inner surface area and high porosity. The first aerogels that were produced were based on inorganic compounds such as silica, but the new aerogels that have emerged in recent years are based on other chemistries, including graphene aerogels [107], titanium aerogels [108], or biopolymers such as polysaccharides [109], starch [110], cellulose [111] chitosan [112] and agar [113]. The biodegradability, biocompatibility, sustainability, and renewability of these aerogels at similar low cost make them ideal for medical, pharmaceutical and food applications. Especially, starch-based aerogel presents excellent mechanical toughness and biodegradability compared with inorganic aerogel. It is still in progress and dominated by fundamental research [114].

Similarly to the silica aerogels, bio-aerogels are produced after a series of steps which are carefully performed under specific conditions. A generally accepted strategy to produce bio-aerogels includes the following procedures: (i) suitable solvents for dissolving biopolymers, (ii) gelation and pore formation, (iii) solvent exchange and pore adjustment, (iv) gel drying processes [115].

The most important area for the application of aerogels are all kinds of thermal and acoustic superinsulation. Although yet discovered in the 1930s, aerogels as a thermal superinsulation material have undergone great progress and changes in recent years. Superinsulation or high-performance insulation is a term which is used for materials with thermal conductivity lower than 0.020 W mK^{-1}. It is well-known in physics that thermal conductivity (λ) is an intrinsic property of a materials ability to conduct heat [116].

On the other hand, superinsulation materials like aerogels are considered to dominate the global market in the upcoming future depending on the high performance for insulation, the extraordinary features and decreasing cost. In fact, the global market of silica-based aerogels is growing at an annual growth rate of 33.8% and it is expected to reach US$1.9 billion by 2022 [117].

The authors are convinced that more novel aerogel types are going to be discovered in the next years, since especially hybrid aerogels appear to enable specific properties and add new functionalities depending on the target application. According to the study of the properties and the application of modified aerogels, it was found that the remarkable textural properties and changeable surface chemistry of aerogel are helpful to increase its absorbent capacity for bacterial and viruses, electrochemical energy storage, air filtrations, etc [115, 118]. Aerogel nanostructures are considered sustainable materials for the fabrication of energy conversion and storage devices.

Finally, due to their high porosity and large specific surface area, aerogels can be used as matrices for various biomedical applications. The use of aerogels in the field of biomaterials is relatively recent. They appear successful as platforms for adsorption and controlled release of various bioactive compounds due to their biocompatibility, and biodegradability [119]. However, the potential of aerogels in the medical field is not sufficiently explored. For example, the use of aerogels in tissue engineering and wound care is not satisfactorily exploited, although it is evident that their properties make them indispensable in regenerative medicine [120]. Future research should be driven with potential applications, determining the properties of the aerogels depending on the synthesis process of silica aerogels.

Conclusion

Aerogels represent an interesting type of porous materials with a variety of high-performance applications thanks to their unique physical properties such as low density, high specific surface area, and porosity. These properties combined with a modifiable surface chemistry (that results from the versatile synthesis), make these materials a suitable candidate for an array of several applications.

Aerogels are mainly obtained by the sol-gel process. There are several alternative techniques, including the self-assembly, emulsion, gelation and drying process.

The properties of the aerogels as well as their microstructure depend strongly on the type of precursors used in synthesis process, and the chemical processes during gel formation and up to the drying step. Drying of the gels may be carried out by supercritical or ambient pressure drying methods, depending on the commercial aerogel application.

Aerogels have applications as material for batteries, solar cells, fuel cells, and supercapacitor applications. Another important application area of aerogels is the acoustic superinsulation originated from the combination of a solid network and nanosized pores filled with air.

Aknowledgment

The authors thank Comecyt - Consejo Mexiquense de Ciencia y Tecnología for catedra support and Centro Conjunto de Investigación en Química Sustentable UAEM-UNAM for the support in the recurrent research stay.

References

[1] Kistler, S. S. *Nature* 127 (1931) 741-741.

[2] Çok, S. S., F. Koç, F. Balkan, N. Gizli, *Ceram. Int.* 45 (2019) 1616–1626.

[3] Xia, T., H. Yang, J. Li, C. Sun, C. Lei, Z. Hu, Y. Zhang, *Ceram. Int.* 45 (2019) 7071–7076.

[4] Sarawade, P. B., J. K. Kim, A. Hilonga, D. V. Quang, H. T. Kim, *Micropor. Mesopor. Mater.* 139 (2011) 138–147.

[5] Maleki, H., L. Durães, A. Portugal, J, *Non-Cryst. Solids* 385 (2014) 55–74.

[6] Gurav, J. L., A. V. Rao, A. P. Rao, D. Y. Nadargi, S. D. Bhagat, *J. Alloy. Compd.* 476 (2009) 397–402.

[7] He, S., Z. Li, X. Shi, H. Yang, L. Gong, X. Cheng, *Adv. Powder Technol.* 26 (2015) 537–541.

[8] Tsou, P. *J Non-Cryst Solids.* 186 (1995) 415–27.

[9] Zu, G., T. Shimizu, K. Kanamori, Y. Zhu, A. Maeno, H. Kaji, J. Shen, K. Nakanishi, *ACS Nano* 12 (2018) 521–532.

[10] Wei, X., C. Ou, S. Fang, X. Zheng, G. Zheng, X. Guan, *Powder Technol.* 345 (2019) 213–222.

[11] Chal, B., B. Yrieix, L. Roiban, K. Masenelli-Varlot, J. M. Chenal, G. Foray, *Energy Build.* 183 (2019) 626–638.

[12] Pinchuk, O. A., F. Dundar, A. Ata, K. Wynne, *J. Int. J. Hydrogen Energy* 37(3) (2012) 2111–2120.

[13] Veres, P., A. M. Lopez-Periago, I. Lazar, J. Saurina, C. Domingo. *Int J Pharm* 496(2) (2015) 360–370.

[14] Gaudio, G. P., G. Auriemma, T. Mencherini, G. D. Porta, E. Reverchon, R. P. Aquino. *J Pharm Sci* 102(1) (2013) 185–194.

[15] Mikkonen, K. S., K. Parikka, A. Ghafar, M. Tenkanen, *Trends Food Sci Technol* 34(2) (2013) 124–136.

[16] Fang, L. X., K. J. Huang, Y. Liu, *Biosens Bioelectron* 71 (2015) 171–178.

[17] Islam, U. M., S. Khan, M. W. Ullah, J. K. Park, Biotechnol J 10(12) (2015) 1847–1861.

[18] Long, J. W., K. E. Swider Lyons, R. M. Stroud, D. R., Electrochem. *Solid-State Lett.* 3 (2000) 453–456.

[19] Ward, D. A., E. I. Ko, J. *Ind. Eng. Chem. Res*. 34 (1995) 421–433.

[20] Wang, C. T., C. L. Wu, I. C. Chen, Y. H. Huang, *Sens. Actuators B* 107 (2005) 402-410.

[21] Perdigoto, M. L. N., R. C. Martins, N. Rocha, M. J. Quina, L. Gando-Ferreira, R. Patrício, L. Durães, J. *Colloid Interface Sci*. 380 (2012) 134–140.

[22] Slowing, I. I., J. L. Vivero-Escoto, C. W. Wu, V. S. Y. Lin, *Adv. Drug Deliv. Rev* 60 (2008) 1278-1288.

[23] Rousset, J. L., A. Boukenter, B. Champagnon J. Dumas, E. Duval, J. F. Quinson, J. Serughetti, *J. Physics: Condensed Matter* 2(42) 1990 8445–8455.

[24] Pajonk, G. M., T. Manzalji, *Catal Letters* 21(3-4) (1993) 361–369.

[25] Alwin, S., X. Sahaya Shajan. *Mater Renew Sustain Energy* (2020) 9:7.

[26] Biener, J., M. Stadermann, M. Suss, M. A. Worsley, M. M. Biener, K. A. Rose, T. F. Baumann, *Energy Environ. Sci*. 4 (2011) 656–667.

[27] Baetensa, R., B. P. Jelle, A. Gustavsend, *Energy and Buildings* 43 (2011) 761–769.

[28] Hüsing, N., U. Schubert, *Angew. Chem. Int. Ed*. 37 (1998) 22 – 45.

[29] Gurav, J. L., I. Jung, H. H. Park, E. S. Kang, D. Y. Nadargi, *J. Nanomaterials*, (2010) doi:10.1155/2010/409310.

[30] Torikai, D., C. K. Suzuki, H. Shimizu, T. Ishizuka, J. Yagi, K. Orii, T. Miyakawa, *J. Non-Cryst. Solids* 179 (1994) 328–334.

[31] Chen, Q., H. Wang, L. Sun. *Materials* 10 (2017) 435.

[32] Saoirse, D., C. S. Pillai. *An Introduction to Sol-Gel Processing for Aerogels, Sol-Gel Materials for Energy, Environment and Electronic Applications* 1-22.

[33] Sinkó, K., *Materials* 3 (2010) 704-740.

[34] Brinker, C. J., G. W. Scherer. *Sol–gel science, the physics and chemistry of sol–gel processing*, Ed. Academic Press, Boston 1990, xiv, 908 pp.

[35] Nadargi, D. Y., R. R. Kalesh, A. V. Rao, *J. Alloy. Compd*. 480(2) (2009) 689–695 (2009).

[36] Montes, G. S., H. Maleki, *Colloidal Metal Oxide Nanoparticles*, Chapter 12 Aerogels and their applications (2019).

[37] Klein, L., M. Aparicio, Andrei Jitianu, *Handbook of Sol-Gel Science and Technology Processing, Characterization and Applications, Second Edition*, Springer, (2005).

[38] Dorcheh, A. S., M. H. Abbasi, *J. Mater. Process. Technol*. 199 (2008) 10–26.

[39] Schubert, U., N. *Hüsing, Synthesis of Inorganic Materials*; Wiley-VCH: Weinheim, Germany, 2005, 192–221.

[40] Jones, S. M., *J. Sol-Gel Sci. Technol*. 44 (2007) 255–258.

[41] Sinkó, K., N. Hüsing, G. Goerigk, H. Peterlik, *Langmuir* 24 (2008) 949–956.

[42] Wagh, P. B., R. Begag, G. M. Pajonk, A. V. Rao, D. Haranath, *Mater. Chem. Phys*. 57(3) (1999) 214–218.

[43] Yu, Y. X., X. Y. Wu, D. Q. Guo, J. Y. Fang, *J Mater Sci* 49 (2014) 7715–22.

[44] Berardi, U. *Aerogel-enhanced Insulation for Building Applications*, ed. Pacheco-Torgal (Nanotechnology in eco-efficient construction), Elsevier, 2019.

[45] Hrubesh, L. W., P. R. Coronado, J. H. Satcher Jr., *J. Non-Cryst. Solids* 285 (2001) 328-332.

[46] Reynolds, J. G., P. R. Coronado, *J. Non-Cryst. Solids* 292 (2001) 127-137.

[47] Chen, N., Q. Pan, *ACS Nano* 7 (8) (2013) 6875–6883.

[48] Smirnova, I., S. Suttiruengwong, W. Arlt, *J. Non-Cryst. Solids* 350 (2004) 54–60.

[49] Cervin, N. T., C. Aulin, P. T. Larsson, L. Wagberg, *Cellulose* 19 (2012) 401–10.

[50] Lee, K. H., S. Y. Kim, K. P. Yoo, K, *J. Non-Cryst. Solids* 186 (1995) 307.

[51] Liu, H., W. Sha, A. T. Cooper, M. Fan, *Colloids Surf. A Physicochem. Eng. Asp*. 347 (1–3) (2009) 38–44.

[52] Pope, E. J. A., J. D. Mackenzie, *J. Non-Cryst. Solids* 87 (1986) 185.

[53] Rao, A. V., D. Haranath, *Micropor Mesopor Mater.* 30 (1999) 267–273.

[54] Cuce, E., P. M. Cuce, C. J. Wood, S. B. Riffat, *Energy and Buildings* 77 (2014a) 28-39.

[55] Wei, G., Y. Liu, X. Zhang, F. Yu, X. Du, *International Journal of Heat and Mass Transfer* 54 (11-12), 2355-2366, 2011.

[56] Thapliyal, P. C., K. Singh, *J. Mater*., Article ID 127049 (2014) 1-10.

[57] Maleki, H., *Chem. Eng. J.* 300 (2016) 98–118.

[58] Brinker, C. J., *J. Non-Cryst Solids*, 100(1–3) (1988) 31–50.

[59] Dervin, S., S. Pillai. Chapter 1. *An Introduction to Sol-Gel Processing for Aerogels*, 2017.

[60] Hæreid, S., J. Anderson, M. A. Einarsrud, D. W. Hua, D. M. Smith, *J. Non-Cryst. Solids* 185(3) (1995) 221–2266.

[61] Hæreid, S., E. Nilsen, M. A. Einarsrud, *J. Non-Cryst. Solids* 204(3) (1996) 228–234.

[62] Maleki, H., L. Duraes, A. Portugal, *Micropor. Mesopor. Mater*. 197 (2014) 116–129.

[63] Leventis, N., A. Palcezer, L. Mccorkle, *J. Sol-Gel Sci. Tech*. 35 (2005) 99–105.

[64] Chen, H. B., E. Hollinger, Y. Z. Wang, D. A. Schiraldi, *Polymer* 55 (2014) 380-384.

[65] Goksu, E. I., M. I. Hoopes, B. A. Nellis, C. Xing, R. Faller, C. W. Frank, S. H. Risbud, J. H. Satcher Jr., M. L. Longo, *Biochim Biophys Acta Biomembr* 1798 (2010) 719–729.

[66] Hench, L. L., J. K. West, *Chem. Rev*. 90 (1990) 33-72.

[67] Kistler, S. S., *J. Phys Chem*. 34 (1932) 52–64.

[68] Cuce, E., P. M. Cuce, C. J. Wood, S. B. Riffat, *Renew Sust Energ Rev* 34 (2014) 273–299.

[69] Karmakar, B., G De, D. Ganguli. *J. Non-Cryst. Solids* 272(2–3) (2000) 119–126.

[70] Tewari, P. H., A. J. Hunt, K. D. Lofftus, *Mater. Lett*. 3(9) (1985) 363–367.

[71] Zuo, L., Y. Zhang, L. Zhang, Y. E. Miao, W. Fan, T. Liu, *Materials* 8 (2015) 6806–6848.

[72] Smith, D. M., R. Deshpande, C. J. Brinke, MRS Online Proc. Library Archive 271 (1992).

[73] Rao, A. P., G. M. Pajonk, A. V. Rao, *J. Mater. Sci*. 40 (2005a) 3481–3489.

[74] He, S., G. Sun, X. Cheng, *Composit. Sci. Technol*. 146 (2017) 91–98.

[75] Vareda, J. P., T. Matias, L. Duraes, *Ceramics Int.* 44 (2018) 17453–17458.

[76] Mathieu, B., S. Blacher, R. Pirard, J. P. Pirard, B. Sahouli, F. Brouers, *J Non-Cryst Solids* 212 (1997) 250–261.

[77] Schwan, M., L. Ratke, *J. Mater. Chem. A* 1 (2013) 13462–13468.

[78] Smith, R. K., P. A. Lewis, P. S. Weiss, *Prog. Surf. Sci.* 75 (1) (2004) 1–68.

[79] Parikh, N. A., D. L. Allara, I. B. Azouz, F. Rondelez, *J. Phys. Chem.* 98 (31) (1994) 7577–7590.

[80] Schwertfeger, F., N. H€using, U. Schubert, *J. Sol-Gel Sci. Technol.* 2 (1–3) (1994) 103–108.

[81] Xiong, X., N. Ji, C. Song, Q. Liu, *Procedia Eng*. 121 (2015) 957 - 960.

[82] Woignier, T., J. Phalippou, M. Prassas, *J. Mater Sci.* 25 (1990) 3118–26.

[83] Perrin, L., A. L. Faivre, S. Calas, T. Woignier, *J Non-Cryst Solids* 333 (2004) 68–72.

[84] Pierre, A. C., G. M. Pajonk, *Chem. Rev.* 2002, 102, 4243–4266.

[85] Pekala, R. W., Low density Resorsinol-Formaldehyde aerogels, in: U.S.D.o. *Energy* (Ed.), 4873 218, US, 1989.

[86] Zhang, R., W. Li, K. X. Li, C. Lu, L. Zhan, L. Ling, *Micropor Mesopor Mater,* 72(1–3) (2004) 167–173.

[87] Hanzawa, Y., K. Kaneko, R. W. Pekala, M. S. Dresselhaus, *Langmuir* 12 (26) (1996) 6167–6169.

[88] Bekyarova, E., K. Kaneko, *Adv. Mater*. 12 (21) (2000) 1625–1628.

[89] Yoldas, B. E., *J. Mater. Sci*., 10 (1975) 1856–1860.

[90] Malinowska, B., J. Walendziewski, D. Robert, J. V. Weber, M. Stolarski, *Int. J. Photoenergy*, 5 (2003) 147–152.

[91] Chaput, F., B. Dunn, P. Fuqua, K. Salloux, *J. Non-Cryst. Solids*, 188 (1995) 188, 11–18.

[92] Zhang, L., T. Wu, H. Na, C. Pan, X. Xu, G. Huang, Y. Liu, Y. Liu, J. Gao. *Ind. Eng. Chem. Res*. 55(23) (2016) 6553–6562.

[93] Xu, Y., K. Sheng, C. Li, G. Shi, *ACS Nano* 4(7) (2010) 4324–4330.

[94] Xu, Y. X., K. X. Sheng, C. Li, G. Q. Shi, *ACS Nano* 4 (2010) 4324-4330.

[95] Arachchige, I. U., S. L. Brock, *J. Am. Chem. Soc*. 129 (2007) 1840.

[96] Ganesan, K., T. Budtova, L. Ratke, P. Gurikov, V. Baudron, I. Preibisch, P. Niemeyer, I. Smirnova, B. Milow, *Materials* 11 (2018) 2144.

[97] Baudron, V., P. Gurikov, I. Smirnova, *J. Colloid Surface A* 566 (2019) 58-69.

[98] Alnaief, M., M. A. Alzaitoun, C. A. García-González, I. Smirnova, *J. Supercrit Fluid* 55(3) (2011) 1118-1123.

[99] Stergar, J., U. Maver, *J. Sol-Gel Sci. Technol*. 77 (2016) 738–752.

[100] Zheng, Q., Y. Tian, F. Ye, Y. Zhou, G. Zhao, *Trends Food Sci Tech*. 99 (2020) 608–620.

[101] Maleki, H., L. Durães, C. A. García-González, P. del Gaudio, A. Portugal, M. Mahmoudi, *Adv Colloid Interface Sci.* 236 (2016) 1-27.

[102] Shaari, N., S. K. Kamarudin, *Int. J. Energy Res*., 43 (2019) 2447–2467.

[103] Maleki, H., N. Hüsing, *Appl. Catal., B*, 221 (2018) 530–555.

[104] Yang, J., Y. Li, Y. Zheng, Y. Xu, Z. Zheng, X. Chen, W. Liu, *Small*, 15 (2019) 1902826.

[105] Maleki, H., Chem. *Eng. J.*, 300 (2016) 98–118.

[106] Xu, X., Q. Zhang, M. Hao, Y. Hu, Z. Lin, L. Peng, T. Wang, X. Ren, C. Wang, Z. Zhao, C. Wan, H. Fei, L. Wang, J. Zhu, H. Sun, W. Chen, T. Du, B. Deng, G. J. Cheng, I. Shakir, C. Dames, T. S. Fisher, X. Zhang, H. Li, Y. Huang, X. Duan, *Science*, 363 (2019) 723–727.

[107] Jiang, Y., S. Chowdhury, R Balasubramanian. *J. Colloid Interf Sci.* 534 (2019) 574–585.

[108] Zhang, C., S. Liua, Y. Qia, F. Cui, X. Yang, *Chem Eng J.* 351 (2018) 825-831.

[109] Batista, M. P., V. S. S. Gonçalves, F. B. Gaspar, I. D. Nogueira, A. A. Matias, P. Gurikov. *Int. J. Biol. Macromol.* (2020), https://doi.org/10.1016/j.ijbiomac. 2020.04.089.

[110] De Marco, I., S. Riemma, R. Iannone, *J. Supercrit Fluid.* 143 (2019) 305–312.

[111] Fan, X., L. Zhang, G. Zhang, Z. Shu, J. Shi, *Carbon* 61 (2013) 423-430.

[112] Cao, N., Q. Lyu, J. Li, Y. Wang, B. Yang, S. Szunerits, *Chem Eng J.* 326 (2017) 17–28.

[113] Chen, L., Y. Li, Q. Du, Z. Wang, Y. Xia, E. Yedinak, *Carbohydr Polym.* 155 (2017) 345–353.

[114] Martins, M., A. A. Barros, S. Quraishi, P. Gurikov, S. P. Raman, I. Smirnova, *J. Supercrit Fluid.* 106 (2015) 152–159.

[115] Zaman, A., F. Huang, M. Jiang, W. Wei, Z. Zhou, *Energy and Built Environment* 1 (2020) 60–76.

[116] Cuce, E., P. M. Cuce, C. J. Wood, S. B. Riffat, *Renew Sust Energ Rev* 34 (2014) 273–299.

[117] Clark, S., *Allied Market Res.* (2016).

[118] Ahmadi, M., A. Madadlou, A. A. Saboury, *Food Chem* 196 (2016) 1016–1022.

[119] Nita, L. E., A. Ghilan, A. G. Rusu, I. Neamtu, A. P. Chiriac, *Pharmaceutics* 12 (2020) 449.

[120] Ahmad, F., Z. Ulker, C. J. Erkey, *Non-Cryst. Solids* 491 (2018) 7–13.

Chapter 6

Aerogels for Biosensors

Salomé Gabriela Galeas Hurtado*

Materials Department, Escuela Politécnica Nacional, Quito, Ecuador

Abstract

Biosensing application require the development of new materials which can meet specific demands. Early detection of diseases and effective treatments represent the main challenges for researchers around the world. This chapter presents an overview of the types of aerogels, their synthesis and characterization methods, as well as their use as biosensors and sensing platforms for wearable devices.

Keywords: aerogel, biosensors, wearable devices

Introduction

A biosensor is an analytical electronic device that integrates a receptor and a transducer to measure biological or chemical parameters. The receptor is the biological recognition element that interacts with the analyte of interest and the response is converted into a detectable electrical or optical signal by a transducer. The most desirable features in a biosensor are: high selectivity, high sensitivity (low detection limit), specificity, reproducibility, stability, wide sensing range, cost-effectiveness and quick response. For wearable devices, biosensors should also be compact, flexible and biocompatible [1–3].

* Corresponding Author's Email: salome.galeas@epn.edu.ec.

In: Aerogels: Properties and Applications
Editor: Alina Iuliana Pruna
ISBN: 978-1-68507-788-4

The manufacture of biosensors that allow better communication between the human body and its environment represents an important advance in medical care, starting from diagnosis in the early stages of the disease and up to monitoring and treatment, thus, improving the patient life quality [4–6]. Many materials have been used as support for sensors and biosensors in the last decades. However, the rapid development of nanotechnology allowed infinite possibilities to obtain accurate, quick responsive and low-cost biomedical devices. Sim et al. [7] fabricated an ultrathin and stretchable device using an indium zinc oxide (IZO) semiconductor nanomembrane, which is worn on human skin to collect signals from human motion, showing the multifunctionality of the IZO nanomembrane to be used as a temperature, strain, and ultraviolet (UV) sensor. Lee et al. [8] developed a stretchable device using graphene doped with gold and combined it with a gold mesh that can be used as a wearable patch for sweat-based diabetes monitoring. Devices based on optical micro and nanofibres (MNF) attracted a considerable scientific interest. In this respect, bioselective molecules inmobilization on the surface of MNF was studied for high sensitivity detection of glucose and Escherichia coli [9], while gold nanoparticles have been applied to silica optical fibers as signal amplifiers for more sensitive and selective cancer biomarkers [10].

Among nanomaterials used for biosensing applications, aerogels have attracted extensive attention recently. Aerogels are solid structures with interesting properties such as ultra-high porosity, low density, high surface area and adjustable structure. Depending on the material they can exhibit diverse characteristics. Silica aerogels show low thermal conductivities, low sound velocities and acoustic impedances, low dielectric constants and refractive indices, high optical transmission [11, 12]. On the other hand, graphene aerogels exibit chemical stability, high electrical and thermal conductivity, unhindered substance diffusion, high elasticity and a strong adsorption capacity [13, 14].

The multifunctional properties of aerogels make them suitable for promising applications like adsorption and catalysis, acoustic and thermal insulation, energy storage and conversion, shock energy absorption, etc. [11, 15, 16]. Thanks to their high internal surface area that is available for the adsorption of guest molecules, aerogels are considered excellent materials for sensing applications [17].

Functionalization of aerogels by chemical doping can result in property improvement. For example, nitrogen doped graphene aerogel offers much better electrochemical performance as compared with graphene aerogel [18].

Tian et al. [19] demonstrated the role of phosphorus doping in enhancing the conductivity of graphene aerogel by improving the electron-donor properties and increasing the density of conducting electrons of carbon at energies near Fermi level, which contributes to the sensing performances of modified electrodes.

The focus of this chapter is to present the most common types of aerogels used for biosensing applications, the synthesis approaches, and the characterization methods. Moreover, this chapter provides a review of aerogel-sensors used for wearable devices, the emerging challenges, and perspectives on the development of aerogel biosensors.

Types of Aerogels

Carbon, graphene and metal oxide-based aerogels are the most common types of aerogels. Nevertheless, in the last three decades, bio-aerogels as polysaccharide-based ones have gained popularity thanks to their non-toxicity and biodegradability, which are desirable aspects in the application asbiosensors and wearable devices. Some of the principal properties and applications of the different types of aerogels are exposed at this point, and the sensing applications are detailed in the next section.

Metal Oxide Aerogels

These aerogels exhibit the common features of aerogels including low density, high porosity, and high specific surface area. Moreover, they also exhibit excellent thermal stability and non-flammability, which allow them to be used as high-temperature catalyst supports. The oxide-based aerogels that have been exhaustively studied are silica, alumina, zirconia, titania and, vanadium oxide [20].

Silica Aerogels

Among the metal oxide-based aerogels, silica aerogels have attracted the most attention as they exhibit a wide variety of unique characteristics such as high porosity, high specific surface area, low dielectric constant and refractive index, high thermal and acoustic insulating properties. Nevertheless, silica aerogels show some disadvantages namely, low compressive and tensile strengths. Consequently, these materials are often functionalized or combined with other metal oxides such as titania, alumina, or iron oxides, which can react with silicon alcoxide precursor to obtain composites with improved properties [11, 12]. The above-mentioned features have made silica aerogels suitable for applications including absorption of organic solvents and oils, heterogeneous catalysis in gas or liquid phase, CO_2 capture and removal of volatile organic compounds from gas streams, fuel storage, ion exchange, soundproof rooms, energy absorber, cosmic particle trapping, dielectrics for integrated circuits, capacitors, thermal insulation, sensing and biosensing.

Carbon Aerogels

Carbon aerogels (CAs) are mesoporous carbon materials with a three-dimensional network structure and attractive physical attributes such as low density, continuous porosity, high specific surface area, controlled pore size, excellent electrical conductivity, chemical and thermal stability, and environmental compatibility [21, 22]. Thus, CAs have attracted increased research interest for supercapacitors, rechargeable batteries, catalyst supports, waste water treatment by adsorbention, thermal insulators, and immobilization of large biomolecules for biosensing [23–25].

Graphene-Based Aerogels

Graphene aerogels (GAs) possess excellent electrocatalytic properties, good chemical and environmental stability, good biocompatibility, low density, large specific area, and abundant reactive sites. Thanks to these characteristics, GAs have been selected in supercapacitor and battery electrodes, hydrogen storage, water treatment, gas sensing and biosensing applications [26–28]. Despite of the above-mentioned properties, graphene tends to aggregate due to strong π– π interlayer interactions, that further affects their properties. The doping of heteroatoms (including nitrogen, phosphorous and boron) into graphene lattice can be applied to redistribute the spin and charge densities and thus, to overcome this limitation and enhance the electrocatalytic properties of graphene [18, 28, 29].

Bioaerogels

Biomedical applications that include drug delivery systems, biosensors and wearable devices require materials that are biocompatible and biodegradable. Biopolymers obtained from natural resources including polysaccharide-based aerogels are widely used in tissue engineering, disease diagnosis, drug delivery, and biosensing [30, 31]. Cellulose represents the most common polysaccharide used to synthetize aerogelsthanks to its high availability and renewability, however chitosan, starch, alginate, pectin, and agar have been studied as well in the past decade [32, 33].

Synthesis Methods

The conventional procedure to prepare aerogels consists of two stages: the first one is the gelation process to form the three-dimensional network from the precursor solution and the secondcorresponds to the drying applied to replace the liquid phase from the gel with air by proper methods. It is also important to note that the precursor content, pH, and gel temperature determine the final morphology and properties of the aerogel [20, 30].

Gelation Process

It is a simple wet chemical synthesis approach that enables the formation of uniform, small particle aerogels. of varied morphologies . This process implies the transition of a system from a liquid “sol” into a solid “gel” phase by a chemical reaction at low temperature [11, 20].

It is worthy to mention that there is a new approach called freeze-drying-induced gelation that skips the formation of the gel. In this method, the aqueous solution of the polymer or the nanoparticles colloid is directly frozen. Such approach is limited to certain materials like poly-vinyl alcohol (PVA), graphene oxide (GO), and metals that include gold, silver, palladium, platinum, etc. [20].

Drying Process

In order to obtain the aerogel the liquid part of the gel must be replaced by air. For this purpose, there are some drying techniques which allow the gel to retain the microstructure:

- Supercritical drying, where the wet gels are dried in an autoclave up at a critical point in order to eliminate the capillary forces. The most common supercritical fluid used is CO_2 [11, 20, 34].
- Ambient pressure prying, where the surface of the wet gel is first chemically modified by replacing the hydrophobic functional groups then subjected to ambient pressure drying. This modification is necessary to avoid an irreversible shrinkage of the gel network during drying [11, 34].
- Freeze-drying, where the pore liquid is frozen and then sublimed in vacuum [11, 34].

Characterization of Aerogels

To characterize the structure of aerogels, different instrumental techniques can be applied. The most common ones are listed below.

X-ray Diffraction (XRD)

The crystalline structure of the aerogels can be studied by X-Ray diffraction. For example, in the work carried out by Ali et al., XRD was used to predict the monocrystalline structure of gold nanoparticles in the as-prepared hybrid aerogel (poly-N-isopropylacrylamide/gold nanoparticles). The obtained diffraction peaks corresponded to crystallographic planes (111), (200), (220), (311) and (222), that are characteristic of gold metal [35].

The XRD pattern is also employed to predict the degree of crystallinity and the crystallite size of the aerogel, by using Segal's empirical method and Scherrer's equation, respectively. For example, Hosseini et al. synthetized a bacterial cellulose (BC)/reduced graphene oxide (rGO) nanocomposite aerogel. They observed that adding GO to the culture medium does not change the diffraction peaks of BC significantly, however, by adding rGO to culture medium, it considerably decreased the degree of crystallinity, thus, influencing the formation of BC [36].

X-ray Photoelectron Spectroscopy (XPS)

This spectroscopic technique provides information on the surface functional groups and elemental states of the aerogel by using the photoelectric effect of elements. Electronic binding energies (bonding states) and chemical components of the aerogels are measured using XPS analysis [14, 37, 38]. For instance, Ai et al. [39] synthetized nitrogen-doped graphene aerogels by using different nitrogen sources. XPS spectroscopy was used for determining the type and content of nitrogen species given that different nitrogen species in the molecular skeleton of graphene can lead to the different chemical or electronic environments of adjacent carbon atoms. XPS spectra showed that nitrogen atoms were doped in the graphene basal plane in three configurations: pyridinic N (binding energy: 398.7 eV), pyrrolic N (binding energy: 399.9 eV), and quaternary N (binding energy: 401.8 eV).

Scanning and Transmission Electron Microscopy (SEM/TEM)

Morphologies of the aerogels can be observed by scanning electron microscopy (SEM). Porous networks can be perceived by this technique. Energy dispersive X-ray spectroscopy (EDX) analysis (also available in SEM microscopes) can be used to identify the elemental composition of the aerogel [14, 40]. Ai et al. employed SEM to observethe hierarchical porous network of the as-prepared aerogel, which was composed of randomly oriented graphene sheets with pore size ranging from sub-micrometers to several micrometers [39].

The microstructure and morphology of aerogels can be analyzed using transmission electron microscopy (TEM), which is a useful technique to identify pores at nanometer scale. High resolution TEM images in Li's et al. research clearly showed nanoclusters with average size of 2 or 3 nm, in conformity with the estimated sizes from Scherrer's equation (XRD patterns) [21]. TEM microscopes are equipped with a selective area electron diffraction mode (SAED). This is an experimental technique to characterize the crystal structure, as can be seen in the work of Ruiyi et al., who synthetized a nitrogen-doped multiple graphene aerogel/gold nanostar composite. In this work, a hexagonal diffraction geometry of graphene was shown on the electron diffraction patterns of the SAED image, indicating alsp that the most inner layer (100) diffraction ring is stronger than the secondary inner layer (110) diffraction ring.

Fourier Transform Infrared Spectroscopy (FTIR)

Fourier Transform Infrared Spectroscopy (FTIR) reveals the representative functional groups in the aerogel by displaying the characteristic peaks of the different vibrations of bonded atoms. For example, FTIR spectra of the aerogel fabricated by Saadatnia et al. confirmed the characteristic bands of polyurethane aerogel at 1725 cm^{-1}, 1600 cm^{-1} and 1315 cm^{-1}, which are due to the C=O stretching vibration, skeleton vibration of C= C in aromatic ring, and C–N stretching vibration, respectively [41].

Krittayavathananon and Sawangphruk synthetized a reduced graphene oxide aerogel labeled with a single strand DNA for biosensing applications [42]. In this work, FTIR spectroscopy was used to evidence the functionalization process of ss-HSDNA molecule on the rGO.

The three-dimensional biochips prepared by Li et al. [21] are composed of a glass slide arrayed with amino aerogel dots. The process to obtain the amino aerogel was sol-gel and included an ionic liquid as the template. In this case, FTIR spectra were employed to verify the removal of the template from the aerogel.

Raman Spectroscopy

The structure and morphology of the aerogels can also be studied by Raman spectroscopy. Raman spectroscopy is considered an important tool, especially for the characterization of carbon and graphene aerogels [39]. Martín-Pacheco et al. [43] characterized graphene quantum dots aerogel by Raman spectroscopy. The D band observed at ~1350 cm^{-1} is associated with the symmetry breaking of the sp2 carbon ring, indicating the presence of sp3 hybridized carbon atoms.

The Raman spectra of the graphene - gold nanohybrid aerogel synthetized by Ruiyi et al. [44] exhibited two typical Raman peaks of graphene materials, namely, D band (1320 cm^{-1}) and G band (1591 cm^{-1}).

Raman spectroscopy was employed by Yan et al. [45] to verify the stability of the porous tungsten disulfide (WS_2) aerogel gas sensor before and after one week of sensing tests at 250°C. The study revealed the Raman spectra remained unchanged, thus, the aerogel presenting high stability.

Nitrogen Porosimetry (Brunauere-Emmette-Teller method - BET)

The specific surface area and porous properties such as average pore size and pore volume of aerogels can be determined with the Brunauer Emmett Teller method by using the adsorption–desorption isotherms of nitrogen at 77 K [45, 46]. Guo et al. [15] fabricated reduced graphene oxide (rGO) – tin oxide (SnO_2) p-n heterojunction aerogels to be used as a sensor for phenol detection. The specific surface area of rGO/SnO_2 composite aerogels was detected by BET analysis, and the results showed that the composites' surface area (278 to 431 m^2/g) was much higher than that of SnO_2 aerogel (51 m^2/g), and therefore, enabled this composite aerogel for sensing applications.

Contact Angle Measurement

Surface hydrophilicity or hydrophobicity of the aerogel surface can be distinguished from the value of the contact angle [11]. Ruiyi et al. [18] studied the effect of nitrogen doping on the surface wettability of graphene aerogel. The contact angle decreased from 114° (pristine graphene aerogel) to 92° (N-doped graphene aerogel), indicating a more hydrophilic surface. The hydrophilicity of an aerogel can also be indicative of its suitability for supercapacitors applications [40]. On the other hand, hydrophobic surfaces are used as coatings for rust prevention, oil removal from water, and chemical separation processes [11].

Biosensing Applications of Aerogels

The aerogel characteristics such as high specific surface area and porous structure make them promising materials for biosensing applications, especially in food detection, and environmental and health monitoring. Among the types of sensors, the electrochemical detection is the most used method for biosensors, due to its fast response and low cost. Some of the principal applications of electrochemical aerogel-based biosensors are discussed next.

Glucose Detection

Diabetes is a disease associated with a metabolic disorder which affects many tissues and organs. Even if there is no known cure for diabetes, an early diagnosis is highly important to prevent, monitor and treat this disease. The glucose level is the main indicator for the diagnosis and control of diabetes [47, 48]. Several materials have been used as platforms for glucose electrochemical sensors. Recent studies have demonstrated excellent electrocatalytic activity of copper/copper oxide aerogels (Cu@Cu_2O) towards glucose detection. Gao et al. [49] used this material as an electrode for a non-enzymatic glucose sensor with high sensitivity in a wide linear range, selectivity, good reproducibility and long-term stability. Similarly, Ling et al. [50] used Cu@Cu_2O aerogels as a platform for non-enzymatic oxidation of glucose. It is important to note that cytotoxicity studies must be performed to determine the suitability of these materials to be used as wearable sensors.

Graphene aerogels are used for glucose detection, as well. Gold functionalized aerogels have demonstrated good results, as they showed high catalytic activity and enhanced electron transfer. They are also biocompatible. Ruiyi and his group [44] synthetized an ultrasensitive sensor using a graphene-gold nanohybrid aerogel. Wang et al. [51] also used gold nanoparticles to functionalize graphene aerogel as a glucose oxidase sensor with a low limit of detection (0.597 μM). Similar results were obtained by the same research group [52] by using platinum instead of gold nanoparticles (LOD = 0.561 μM). Finally, it was reported that an amphiphilic aerogel can act as a glucose oxidase immobilization substrate. Ma et al. [53] developed a low-cost poly(vinyl alcohol)/maleic acid aerogel to detect glucose in whole blood and sweat with a low detection limit (11.4 μM).

Dopamine Detection

Dopamine is a neurotransmitter in the central nervous system of human beings and animals. Dopamine concentration in human brain is related to emotions, and attention, and it has a critical role in renal system and metabolism. Abnormal DA levels in the brain are associated to Parkinson, Alzheimer and psychiatric disorders like schizophrenia and depression [38, 39]. Several efforts have been made in the last decades to fabricate effective aerogel-based biosensors to detect dopamine levels in human brain.

Carbon and graphene-based aerogels have been reported for dopamine detection with notable results. As mentioned before, graphene displays excellent conductive properties, however, a functionalization is often needed for electrochemical applications. Ai et al. [39] synthetized nitrogen-doped graphene aerogels with three-dimensional network structure by using different organic amines as nitrogen sources and various reducing agents for the graphene oxide. The biosensor with aniline as nitrogen source showed the best mechanical strength and selectivity towards dopamine in presence of uric acid and ascorbic acid. The macroporous carbon aerogel microelectrode fabricated by Ding et al. [38] displayed excellent electrical conductivity and a rapid response of dopamine detection with a LOD of 30 nM (S/N = 3). Dopamine determination in human serum samples verified a high selectivity in the presence of higher concentration of uric and ascorbic acid. Other researchers have achieved good results, as well. For example, Li et al. [54] used palladium, gold, nitrogen and sulphur to functionalize graphene aerogel and were able to detect as low as 0,36 nM of dopamine. Similar results were obtained by using histidine to functionalize graphene quantum dot graphene aerogel (0.29 nM) [55], by modifying carbon aerogel electrode with molecularly imprinted polypyrrole (0.4 nM) [56], or by incorporating chitosan to carbon nanotubes aerogel (0.3 nM) [57].

Antigen and Cancer Cell Detection

Diagnosis of cancer and other diseases is important in preventing complications and to offer an early treatment for patients. Electrochemical biosensors have been developed with good results.

Aerogel-based biosensors allow a sensitive and selective alternative to traditional sensors, due to their great characteristics. In this respect, Chen et al. [58] synthetized a three-dimensional nitrogen-doped graphene aerogel with Fe_3O_4 nanoparticles as a highly sensitive platform for deoxyribonucleic acid (DNA) detection, thus, assisting early diagnosis of genetic diseases and mutations. Other works regarding DNA detection have been developed by Ruiyi et al. for molecule diagnosis, bioanalysis and catalysis. Thionin-functionalized multiple graphene aerogel and gold nanostars [59] and nitrogen-doped multiple graphene aerogel/gold nanostar [18] were used for direct sensing of double-stranded DNA (dsDNA). Yuanfeng et al. [60] fabricated an ultrahigh sensitive, specific and stable sensor for circulating tumor DNA (ctDNA).

Jia et al. [26] immobilized β-cyclodextrin polymer (Pβ-CD) in graphene aerogel for highly sensitive sensing of the carbohydrate antigen 15–3 (CA15–3) cancer marker. In other work [61], a 3D nano PEDOT (poly 3,4-ethylenedioxythiophene) graphene aerogel with gold nanoparticles was designed as a sensing platform for detecting prostate specific antigen. Yang et al. have also developed a sensor for real sample analysis of prostate specific antigen (PSA) by using palladium nanoparticles to functionalize graphene aerogel supported Fe_3O_4.

Hydrogen Peroxide Detection

High levels of hydrogen peroxide (H_2O_2) in human blood are associated with cancer and progressive neurodegenerative diseases. For this reason, several sensing platforms have been developed for an early detection of H_2O_2. Metal nanoparticles and oxides have been combined with carbon and graphene aerogels to achieve highly sensitive and biocompatible sensors for the detection of trace amounts of H_2O_2 released by living cells.

Xie et al. [62] assembled gold nanoparticles in a nitrogen doped carbon aerogel to fabricate an efficient nonenzymatic biosensor for H_2O_2 released from living cells. Zhao et al. [63] have synthetized 3D graphene aerogel decorated with gold nanoparticles to immobilize Cytochrome c (Cyt c) for H_2O_2 detection. A facile fabrication approach of graphene aerogel/platinum nanoparticle nanocomposites was reported by Yan et al. [64]. This biosensor exhibited a good sensitivity, reproducibility, and long-term stability.

Contaminant Detection

The rapid growth of urban, agricultural, and industrial activity during the last decades has resulted in adverse impacts on the environment. Among the pollutants present in wastewater, it is worthy to mention the organic pollutants such as dyes, phenols, pesticides, insecticides, and fertilizer residues.

Carbon aerogel (CA) was modified with molecularly imprinted TiO_2 by Zhang et al. (136) for atrazine monitoring in water. The electrode presented high sensitivity and selectivity while used in co-existing systems containing various interferents in concentration of 25 times that of atrazine. Baia et al. [65] fabricated porous nanoarchitectures based on TiO_2 aerogels and Au colloidal particles to be used as sensors in surface-enhanced Raman spectroscopy (SERS). These sensors were able to detect contaminant species as thioacetamide and dye molecules off and under resonant excitation. Liu et al. [66] developed an acetylcholinesterase/platinum–carbon aerogels composite biosensor to detect organophosphorous pesticides (OPs). Sensing tests were performed to detect methamidophos and monocrotophos and the response was fast, accurate and sensitive with good reproducibility and acceptable stability. Kehrle et al. [67] deposited silicon nanocrystals into silica aerogels to obtain photoluminescent silica hybrids. The high porosity, transparency, and superhydrophobic character of the synthetized aerogel make this material a potential sensing platform for photoluminescence detection of minimal concentrations of 3-nitrotoluene in water.

These applications are not limited to contaminants in water. Hongxia et al. [68] functionalized graphene aerogel with glutamic acid for acetamiprid detection in vegetables with high sensitivity and selectivity. Hu et al. [69] fabricated a fluorescence microfluidic sensor based on quantum dots (QDs) aerogel and acetylcholinesterase enzyme (AChE). This biosensor was used for accurate detection of organophosphate pesticides (OPs) in spiked fruit samples with a low detection limit and a wide linear range.

Table 1 summarizes the research works of aerogel-based biosensors and their performance in the detection of different analytes.

Sensors for Wearable Devices

Wearable sensors have attracted great interest due to their potential applications in health monitoring. Aerogels are considered promising materials for application in wearable devices thanks to tTheir inherent propertiesincluding low density, high porosity, and large specific surface. Many efforts have been performed to develop aerogel-based sensors to detect human bio-signals such as pulse, speech recognition, motion, etc. [89–91]. Table 2 shows pressure, compression, and strain sensors with potential applications in wearable sensors.

Table 1. Electrochemical aerogel-based biosensors

Material	Analyte	Linear Range	Limit of Detection	Sensitivity	Selectivity	Recovery	Ref.
β-cyclodextrin and graphene aerogel	Antigen	10^{-4} - 100 U mL^{-1}	0.03 mU mL^{-1}	1.14 μA (U $mL^{-1})^{-1}$ cm^{-2}	-------	-------	[26]
Polypyrrole Zeolitic imidazolate framework/graphene aerogel	Dichlorophenol	$3*10^{-10}$ - 10^{-5} M	0.1 nM	-------	Good	95.96 - 104.6 (lake water)	[70]
Nitrogen-doped multiple graphene aerogel/gold nanostar	DNA	10^{-21} - 10^{-16} g mL^{-1}	$3.9*10^{-22}$ g mL^{-1}	-------	High	-------	[18]
Glassy carbon electrode modified with thionin-functionalized multiple graphene aerogel and gold nanostars	DNA	10^{-4} - 10 ng ·mL^{-1}	39 fg mL^{-1}	-------	-------	-------	[59]
Urchin-like gold nanocrystal-multiple graphene aerogel	DNA	0.1 - 10^{6} fM	0.031 fM	-------	Good	95.8 - 104.8 (blood)	[60]
Reduced carbon aerogel	Dopamine	0.2 - 90 μmol L^{-1}	30 nM	-------	-------	-------	[38]
Palladium and gold nanoalloys/nitrogen and sulphur-functionalized multiple graphene aerogel	Dopamine	0,001-100 μM	0,00036 μM	-------	-------	-------	[54]
Histidine-functionalized graphene quantum dot-graphene micro-aerogel	Dopamine	10^{-9} – $8*10^{-5}$ M	$2.9*10^{-10}$ M	-------	Good	-------	[55]
Carbon aerogel electrode with molecularly imprinted polypyrrole	Dopamine	0.007 - 35 μM	0.0004 μM	-------	Good	97.23 - 102 (serum)	[56]
Fe_2O_3 Nanoparticles on Graphene-Polyimide Aerogel	Dopamine	5 – $5*10^{-4}$ M	$10.9*10^{-4}$ M	-------	High	-------	[71]
Chitosan/Multiwalled Carbon Nanotubes Aerogel	Dopamine glucose	0,017 – 0.3 μM -------	0.3 nM -------	12.8 μA mM^{-1} 8.6 μA mM^{-1}	------- -------	------- -------	[57]

Material	Analyte	Linear Range	Limit of Detection	Sensitivity	Selectivity	Recovery, %	Ref.
Glassy carbon electrode modified with a hybrid material consisting of graphene aerogel and octadecylamine-functionalized carbon quantum dots	Acetaminophen	0.001–10 μM	0.38 nM	-------	-------	96.4 - 104.4	[72]
Aptamer and G-quadruplex DNAzyme modified 3D graphene aerogel composite	Drug (Streptomycin)	$1.4*10^{-12}$ - $2.8*10^{-9}$ mol L^{-1}	$9.2*10^{-14}$ mol L^{-1}	-------	Good	96.7 - 101 (cucumber) 98.0 - 103.3 (milk)	[73]
Nickel nitride nanoparticles on conductive 3D graphene aerogels	Glucose H_2O_2	0.1 - 7645.3 μM 0.005 - 75.13 mM	0.04 μM 1.80 μM	0.9 mA mM^{-1} cm^{-2} 0.1 mA mM^{-1} cm^{-2}	Good	94.6 - 104 (serum)	[74]
Three-dimensional Cu@Cu_2O aerogels	Glucose: alkaline pH Neutral pH	 0.001-17.1 mM 0.1 - 10 mM	 0.6 μM 54 μM	 195 mA $M^{-1}cm^{-2}$ 12 mA $M^{-1}cm^{-2}$	 ------- -------	 ------- -------	[49]
Porous Cu@Cu_2O Aerogel	Glucose	0.05 - 8.0 mM	15 μM	-------	-------	-------	[50]
Polymer poly(vinyl alcohol) (PVA)/maleic acid (MA)	Glucose	-------	11.4 μM	-------	-------	94.8 - 105.3	[53]
Graphene-gold	Glucose	0.01 - 16 mM	0.004 mM	-------	-------	95.7 - 104.2	[44]
Tungsten oxide aerogel	Glucose	0.01 - 1.0 mM	10 μM	0.27 μA $mmol^{-1}$ cm^{-2}	-------	-------	[75]
Graphene aerogel/platinum nanoparticle	Glucose	50 - 500 μmol L^{-1}	0.561 μmol L^{-1}	84.92 μA mM^{-1} · cm^{-2}	Good	-------	[51]
3D Porous Graphene Aerogel and glucose oxidase	Glucose	1 - 18 mM	0.87 mM	-------	-------	90 - 106	[76]
Graphene aerogel/gold nanoparticle	Glucose Oxidase	50 - 450 μmol L^{-1}	0.597 μmol L^{-1}	257.60 μA mM^{-1} cm^{-2}	Good	-------	[52]

Table 1. (Continued)

Material	Analyte	Linear Range	Limit of Detection	Sensitivity	Selectivity	Recovery	Ref.
Dumbbell-Like Gold Nanoparticles with a Porous Nitrogen Doped Carbon Aerogel	H_2O_2	0.05 μM - 20 mM	0.015 μM	307.1 μA (mM $cm^2)^{-1}$	-------	-------	[62]
Graphene aerogel / Pt nanoparticle	H_2O_2	1 - 2785 μmol L^{-1}	0.33 μmol · L^{-1}	138.23 μAmM^{-1} cm^{-2}	-------	-------	[64]
Graphene/gold nanoparticles	H_2O_2	10 –740 μM 0.74 – 9.74 mM	1.1 μM	352 $\mu A \cdot mM^{-1} cm^{-2}$ 128 $\mu A \cdot mM^{-1} cm^{-2}$	Good	100.8 - 103	[63]
MoS_2/graphene	H_2O_2 and glucose	2 - 20 mM	0.29 mM	3.36 μA mM^{-1}	Good	-------	[77]
Molecularly imprinted TiO_2/carbon aerogel	Herbicide (Atrazine)	0.12 - 93 μM	0.008 μM	-------	Good	98.29 – 102 (surface water)	[78]
Oligonucleotide-Au functionalized nanosilica and graphene oxide aerogel	Insuline	0.007 - 5 nmoL L^{-1}	1.6 pmoL L^{-1}	-------	-------	98 - 103.3	[13]
Folic acid and octadecylamine functionalized graphene aerogel microspheres	Liver cancer cells	5 - 10^5 cell mL^{-1}	5 cells mL^{-1}	-------	-------	-------	[79]
Acetylcholinesterase based on platinum–carbon aerogels	Methamidophos Monocrotophos	10^{-11} - 10^{-6} M	0.3 pM 2.7 pM	-------	-------	-------	[66]
Quantum dots (QDs) aerogel and acetylcholinesterase enzyme (AChE)	Organophosphorous pesticides (OPs)	10^{-5} - 10^{-12} M	0.38 pM	-------	-------	98 - 110	[69]
MoS_2/carbon aerogel composites	Platelet-derived growth factor BB	0.001 - 10 nM	0.3 pM	High	Good	-------	[22]
3D graphene aerogel	Oxalic acid	4 to 100 μM	0.8 μM	677.98 mA/mM	-------	-------	[81]

Material	Analyte	Linear Range	Limit of Detection	Sensitivity	Selectivity	Recovery	Ref.
Polydopamine-Capped Bimetallic Au Pt Hydrogel	Paraoxonethyl	0.5 - 1000 ng L^{-1}	0.185 ng L^{-1}	-------	-------	Tap water: 102.6 - 103.46 Lake water: 94.63 - 94.86	[80]
Gold-Pd / dodecylamine-functionalized graphene QDs - graphene aerogel	Peanut allergen Ara h 1	$10^{-22} – 10^{-17}$ M	$4.7*10^{-23}$ M (S/N = 3)	-------	-------	-------	[82]
Graphene aerogel-glutamic acid functionalized graphene QDs /Au nanostars	Pesticid (acetamiprid)	0.001 - 1000 pM	0.37 fM	-------	-------	97.6 to 103.1%	[68]
3D Pd-Graphene Aerogel-Supported Fe_3O_4	Prostate specific antigen	0.0001 - 50 ng mL^{-1}	0.056 pg mL^{-1}	-------	-------	-------	[83]
Au - poly(3,4-ethylene dioxythiophene)- graphene aerogel	Prostate specific antigen	0.0001 - 50 ng mL^{-1}	0.03 pg mL^{-1}	-------	-------	-------	[61]
Cotton- nanocellulose aerogel	Protease	-------	0.05 - 0.25 U mL^{-1}	-------	-------	-------	[84]
Peptid-cellulosic aerogel	Protease	-------	0.13 U/mL	-------	-------	-------	[85]
Au / Graphene Aerogel	Cytochrome c	5 - 150 µmol L^{-1}	1.75 µmol L^{-1}	-------	Good	-------	[86]
Reduced GO aerogel labeled with a single strand DNA	Toxin (Aflatoxin B1)	0.1 - 70 ng mL^{-1}	0.04 ng mL^{-1}	-------	-------	-------	[42]
3D B and N co-doped graphene hydrogels	Toxin (microcystin-LR)	-------	0.03 pM	-------	-------	91 - 104	[87]
CNTs/MoS_2 aerogel modified with H7-polyclonal antibodies	Virus (avian influenza H7)	1-25 ng mL^{-1}	0.43 ng mL^{-1}	-------	-------	-------	[88]

Table 2. Aerogel-based sensors for wearable devices

Material	Type of sensor	Application	Sensing range	Sensitivity	Stability	Ref
Cellulose nanofiber aerogel	Compression and humidity	Pedometer	--------	--------	--------	[92]
3D Graphene–poly(vinyl alcohol) composite aerogels	Motion	Monitoring of motions of human joints with different orientations	--------	0.34 kPa^{-1}	over 5000 cycles	[93]
Polyurethane aerogel	Motion	Biomechanical sensor (arm motions and human walking monitoring)	--------	30 mV/°	--------	[41]
Poly (p-phenylene benzobisoxazole) aerogel	Motion	Self-powered wearable device	--------	--------	--------	[94]
Carbon aerogel	Pressure	Monitoring of human jugular venous pulse, current change when speaking different words, joint movements (finger and wrist bending) - Wearable devices and electronic skins.	Pressure: 0–16.89 kPa	--------	--------	[91]
Polyimide/Carbon Nanotube Composite (PI/CNT)	Pressure	Human motion detection including blowing, carotid pulse, vocal cord vibration and movement (walking and finger movement)	80% strain, 61 kPa	11.28 kPa^{-1}	1000 cycles (for 30% strain at a strain rate of 10 mm min^{-1})	[90]
Graphene-Based Aerogels	Pressure	Great potential for wireless pressure sensors for human motion	--------	0.6053 kPa^{-1}	--------	[89]

Material	Type of sensor	Application	Sensing range	Sensitivity	Stability	Ref
Multi-arch structured all-carbon aerogel	Pressure	Wearable pressure sensor: Wrist pulse rating, throat vibration and muscular motion	9.13 - 7.29 kPa^{-1} 1.43 - 13.56 kPa	--------	--------	[95]
Cellulose Nanofiber – GO / glucose-kaolin carbon aerogel	Pressure	Detection of human joint actions (elbow and finger bendings)	--------	--------	--------	[96]
Carbon Nanotube/ Chitosan aerogel	Pressure	Wearable piezoresistive sensors (human biological signals)	0–2.0 kPa	4.45 kPa^{-1}	--------	[97]
MXene/Reduced Graphene Oxide aergogel	Pressure	Human pulse monitoring	--------	22.56 kPa^{-1}	over 10 000 cycles	[98]
Carbon aerogel	Pressure	Human activities and pulse vibration	--------	1.841 kPa^{-1}	--------	[40]
Cu nanowire aerogel	Pressure	Potential application as electronic skins, wearable electronic devices	--------	0.02 - 0.7 kPa^{-1}	--------	[99]
Cu nanowires/ Polyvinyl alcohol aerogel	Pressure	Medical applications (orthopedist)	--------	--------	10,000 cycles (10.5 kPa, 10 Hz)	[100]
Alginate/N-doped CNT-supported Ag aerogel	Pressure	Flexible touch displays and prosthetic skins	--------	--------	--------	[101]
MXene Sphere/Reduced Graphene Aerogel	Pressure	Soft robotics and wearable devices for monitoring of human activities	0–10 kPa	0.159 kPa^{-1} that is able to sustain from 0 to 6 kPa	6000 cycles	[102]
Graphene/Polyorganosiloxane aerogel	Pressure Strain Temperature	Wearable devices for monitoring of human activities and body temperature	0.01 - 110 kPa 0.1–80% 20–100°C	--------	--------	[103]

Table 2. (Continued)

Material	Type of sensor	Application	Sensing range	Sensitivity	Stability	Ref
Carbon Aerogel/Iron Rubber	Pressure and magnetism sensing	Monitoring of human breath, blood repulse, thoracic volume	0-2 kPa (pressure) 0-500 Oe (magnetic field)	0.004 kPa^{-1} at compressive stress beyond 6 kPa	--------	[104]
Carbon Aerogels	Pressure Strain	Wearable electronics and electronic skins for detecting pronunciation, human face expressions, jugular venous pulse, pulse- fluctuant signal of the arm, joint movements	0–10 kPa (compression) 0–95% (strain)	--------	over 100 cycles (99% strain) at least 100 000 cycles (long-term compression at 50% strain)	[105]
Polyaniline/bacterial cellulose/chitosan/ polydimethylsiloxane aerogel	Pressure Strain	Wearable devices for detecting human motions, joint movement to speech recognition	0.159 kPa^{-1} to 6 kPa^{-1}	High sensitivity	--------	[106]
Bacterial cellulose/MWCNT nanocomposite aerogel	Strain	Strain sensor in human detection motion	--------	--------	--------	[107]
CNT/polyvinyl alcohol aerogel	Strain	Wearable electronics	--------	61 kPa^{-1}	--------	[108]
Graphene/Polydimethylsiloxane : silicone rubber aerogel	Strain	Wearable electronics	--------		--------	[109]
Vanadium nitride/CNT aerogel	Strain	Monitoring of body motion, breath, physical signals, speaking, medical care signals	--------	334 kPa^{-1} for pressures between 3.3 and 6.4 kPa	--------	[110]
Reduced graphene oxide and PEDOT:PSS aerogel	Strain and temperature	Monitorionf of human joint motions	30°C to 50°C	--------	--------	[111]

Material	Type of sensor	Application	Sensing range	Sensitivity	Stability	Ref
Polyimide-derived carbon nanofiber aerogel	Stress	Wearable devices	--------	609 kPa^{-1} in the range of 6.4–10 kPa	plastic deformation of 1.3% after 1000 compressive cycles compressive strength and modulus are remained over 76% after 1000 compressive cycles	[112]
N, B, S atoms co-doped graphene aerogel	Stress	Portable electronics to detect human motion: finger bending, foot step.	--------	--------	94.24% at 5000 cycles	[113]

Gas Sensing

Even if gas sensors cannot always be cataloged as biosensors, it is worthy to discuss some gas sensing applications of aerogels, as they are directly related to human health and environmental concern.

During the last decades aerogels have attracted interest in gas sensing thanks to their high sensitivity to gases and water vapors in air [114]. Ayers and Hunt [115] demonstrated that silica aerogel can act as an excellent photoluminescent sensor element for optical oxygen sensors. Plata et al. developed an aerogel-platform gas sensors that responded to changes in oxygen concentration by entrapping probe species within the aerogel matrix. The probes were: tris(2,2'-bipyridyl) ruthenium(II), ruthenium(II) 4,7-diphenyl-1,10-phenanthroline, and platinum octaethyl-porphine, that are typically used for luminescent detection.

Ma et al. [116] developed a simple, environmentally friendly and low-cost method to obtain fluorescent carbon dots - silica aerogels for detecting aniline gas, and their results indicated a high selectivity and good sensitivity.

Humidity sensors were fabricated by Wang et al. [114] by using thin films of silica nanoparticles aerogel on gold-patterned alumina substrates. Impedance and capacitance measurements of the sensor indicated sensitivity to moisture at 40% or higher relative humidity and presented operation stability and reproducibility. Comparable results were achieved by using a hybrid aerogel sensor of poly-N-isopropylacrylamide with gold nanoparticles [35] and resorcinol–formaldehyde aerogels sensors [117].

Polyethylenimine-capped carbon quantum dots were used by Wang et al. [118] to functionalize silica aerogels for the detection of NO_2 gas. The obtained aerogel sensors had abundant pores and exhibited strong fluorescence activity which could be sensitively and selectively quenched by NO_2. Other researchers have reported excellent results in the NO_2 sensing by using three-dimensional molybdenum disulfide aerogel [119, 120], graphene aerogel decorated with SnO_2 nanoparticles [121] and nanocrystals [122], silver modified 3D reduced graphene oxide aerogel [123], boron doped graphene aerogel [29], and porous tungsten disulfide - graphene aerogel [45, 124].

Yang et al. [16] deposited TiO_2 nanoparticles over silica aerogel films by atomic layer deposition to enhance the performance for H_2S gas sensing. The improvement could be explained by the larger specific surface area of the aerogel, the minor grain size, together with the lattice mismatch between TiO_2 nanoparticles and SiO_2 aerogels, which produced more active sites and lattice defects, and subsequent increase of oxygen vacancy, thus, enhancing the gas sensitivity. Similarly, Shu et al. [125] developed a H_2S sensor by using copper oxide and polypyrrole conductive aerogel doped with graphene oxide.

A sensitive, selective, and reversible fluorescence sensor for SO_2 gas has been synthetized by You et al. by using perovskite quantum dots encapsulated into superhydrophobic silica aerogels and a 3D-printed gas flow-through cell.

Table 3 resumes the principal research works related to development of aerogel-based sensors for gases.

Table 3. Aerogel-based sensors for gas detection

Material	Analyte	Range	Limit of detection	Reference
Silicon Nanocrystal – Silica Aerogel	3-Nitrotoluene	--------	270 μM	[67]
Thiourea-treated graphene aerogel	Ammonia gas	0.02 to 85 ppm	100 ppb	[126]
Carbon dots - silica aerogel	Aniline gas	96 to 960 ppm	77.6 ppm	[116]
Copper oxide - Polypyrrole Conductive Aerogel Doped with Graphene Oxide	Hydrogen sulfide	15 μM to 250 μM	8.6 μM	[125]
TiO2/SiO2 aerogel	Hydrogen sulfide	--------	0,5 ppm	[16]
Poly-N-isopropylacrylamide – gold nanoparticles hybrid aerogel	Humidity	30 to 60% HR	--------	[35]
Resorcinol–formaldehyde aerogels	Humidity	15 to 95% RH at 30°C	--------	[117]
Silica aerogel	Humidity	20 to 90%	--------	[114]
Three-dimensional tungsten disulfide and reduced graphene oxide aerogel	Nitrite	0.01 to 130 μM	3 nM	[127]
Platinum and Tin Oxide nanoparticles – Carbon nanotubes aerogel	Nitrogen dioxide	1 to 5 ppm	--------	[37]
3D MoS2 Aerogel	Nitrogen dioxide	--------	28 ppb	[120]
MoS2/Graphene Hybrid Aerogel	Nitrogen dioxide	--------	50 ppb	[119]
Three-dimensional (3D) graphene aerogel-supported SnO2 (SGA) nanoparticles (NPs)	Nitrogen dioxide	10 to 200 ppm	--------	[121]
3D Ag-rGO	Nitrogen dioxide	80 to 5000 ppb	6.9 ppb	[123]

Table 3. (Continued)

Material	Analyte	Range	Limit of detection	Reference
(3D) SnO2/rGO composites aerogel	Nitrogen dioxide	14 to 110 ppm at 55°C	2 ppm	[128]
Boron doped graphene aerogel	Nitrogen dioxide	--------	7 ppb	[29]
Carbon Quantum Dot functionalized silica aerogels	Nitrogen dioxide	--------	250 ppb	[118]
Porous tungsten disulfide/Graphene aerogel	Nitrogen dioxide	--------	8 ppb for NO2 60 ppm for H2 13 ppm for NH3	[45]
Tungsten disulfide/graphene aerogel	Nitrogen dioxide, oxygen, hydrogen and humidity	--------	10–15 ppb	[124]
Reduced graphene oxide (rGO)-tin oxide (SnO2) p-n heterojunction aerogels	Phenol	10-80 ppb	5 ppb	[15]
Silica aerogels encapsulated fluorescent perovskite quantum dots	Sulphur dioxide	0–10 ppm	155 ppb	[129]

Conclusion

Significant advances in nanoscience and nanotechnology in the last decades enabled the development of materials with versatile properties for numerous applications. Medicine, as one of the most important sciences in human life, has benefited greatly of new material development. Early diagnosis, monitoring, and treatment of diseases, as well as food contamination and environmental, can be achieved thanks to the utilization of biosensors, with the purpose of improving the patient's life quality and extend the human health span.

Great efforts have been made to design sensing platforms that meet the requirements for accurate disease detection. However, traditional biosensors are found expensive, and time consuming. In this respect, aerogels are attracted special attention because of their amazing combination of properties, which enables them to be used as sensors and biosensors. Aerogel-based biosensors show high sensitivity, selectivity, low detection limit, wide linear range, fast response and recovery, thanks to their high porosity, specific surface area, and low density.

As shown in this chapter, many research groups are currently working on the design of aerogel-based biosensors and have achieved great results. However, it is important to notice some challenges which are yet to solve, for example, the biocompatibility and toxicity of wearable devices over time. Even if toxicity studies have been performed in some cases, it is necessary to carry out studies over long time periods of exposure of wearable devices. These devices are electronic ones, therefore, the radiation exposure is one of the typical concerns, especially if they are wirelessly connected to other devices such as cellphones or computers. Several studies have already identified health risks associated to cell phone radiation exposure.

Biomedical application of aerogels represent a field in continuous expansion, as they have a huge potential to be used in drug delivery systems, biocomputers and multifunctional wearable devices. Applications are limited only by researchers' imagination.

References

[1] Arya, A., A. Gangwar, and A. Kumar, *Biosensors in animal biotechnology*. Elsevier Inc., 2019.

[2] Naghdi, T., et al., "Chitin Nanofiber Paper toward Optical (Bio)sensing Applications," *ACS Appl. Mater. Interfaces*, vol. 12, no. 13, pp. 15538–15552, 2020, doi: 10.1021/acsami.9b23487.

[3] Prakash, S., T. Chakrabarty, A. K. Singh, and V. K. Shahi, "Polymer thin films embedded with metal nanoparticles for electrochemical biosensors applications," *Biosens. Bioelectron.*, vol. 41, no. 1, pp. 43–53, 2013, doi: 10.1016/j.bios.2012.09.031.

[4] Doucey, M. A., and S. Carrara, "Nanowire Sensors in Cancer," *Trends Biotechnol.*, vol. 37, no. 1, pp. 86–99, 2019, doi: 10.1016/j.tibtech.2018.07.014.

[5] Chung, S., P. Chandra, J. P. Koo, and Y. B. Shim, "Development of a bifunctional nanobiosensor for screening and detection of chemokine ligand in colorectal cancer cell line," *Biosens. Bioelectron.*, vol. 100, no. July 2017, pp. 396–403, 2018, doi: 10.1016/j.bios.2017.09.031.

[6] Li, H., et al., "Single-molecule detection of biomarker and localized cellular photothermal therapy using an optical microfiber with nanointerface," *Sci. Adv.*, vol. 5, no. 12, pp. 1–10, 2019, doi: 10.1126/sciadv.aax4659.

[7] Sim, K., et al., "Metal oxide semiconductor nanomembrane–based soft unnoticeable multifunctional electronics for wearable human-machine interfaces," *Sci. Adv.*, vol. 5, no. 8, pp. 1–11, 2019, doi: 10.1126/sciadv.aav9653.

[8] Lee, H., et al., "A graphene-based electrochemical device with thermoresponsive microneedles for diabetes monitoring and therapy," *Nat. Nanotechnol.*, vol. 11, no. 6, pp. 566–572, 2016, doi: 10.1038/nnano.2016.38.

[9] Li, Y., et al., "Immobilized optical fiber microprobe for selective and high sensitive glucose detection," *Sensors Actuators, B Chem.*, vol. 255, pp. 3004–3010, 2018, doi: 10.1016/j.snb.2017.09.123.

[10] Zhang, L., Y. Tang, and L. Tong, "Micro-/Nanofiber Optics: Merging Photonics and Material Science on Nanoscale for Advanced Sensing Technology," *iScience*, vol. 23, no. 1, p. 100810, 2020, doi: 10.1016/j.isci.2019.100810.

[11] Gurav, J. L., I. K. Jung, H. H. Park, E. S. Kang, and D. Y. Nadargi, "Silica aerogel: Synthesis and applications," *J. Nanomater.*, vol. 2010, no. May 2014, 2010, doi: 10.1155/2010/409310.

[12] Amonette, J. E., and J. Matyáš, "Functionalized silica aerogels for gas-phase purification, sensing, and catalysis: A review," *Microporous Mesoporous Mater.*, vol. 250, pp. 100–119, 2017, doi: 10.1016/j.micromeso.2017.04.055.

[13] Sun, Y., et al., "A highly selective and sensitive detection of insulin with chemiluminescence biosensor based on aptamer and oligonucleotide-AuNPs functionalized nanosilica @ graphene oxide aerogel," *Anal. Chim. Acta*, vol. 1089, pp. 152–164, 2019, doi: 10.1016/j.aca.2019.09.004.

[14] Cai, Z. X., X. H. Song, Y. Y. Chen, Y. R. Wang, and X. Chen, "3D nitrogen-doped graphene aerogel: A low-cost, facile prepared direct electrode for H2O2 sensing," *Sensors Actuators, B Chem.*, vol. 222, pp. 567–573, 2016, doi: 10.1016/j.snb.2015.08.094.

[15] Guo, D., et al., "Reduced-graphene-oxide/metal-oxide p-n heterojunction aerogels as efficient 3D sensing frameworks for phenol detection," *Carbon N. Y.*, vol. 99, pp. 571–578, 2016, doi: 10.1016/j.carbon.2015.12.074.

[16] Yang, F., et al., "Three-dimensional TiO2/SiO2 composite aerogel films via atomic layer deposition with enhanced H2S gas sensing performance," *Ceram. Int.*, vol. 44, no. 1, pp. 1078–1085, 2018, doi: 10.1016/j.ceramint.2017.10.052.

[17] Dolai, S., S. K. Bhunia, and R. Jelinek, "Carbon-dot-aerogel sensor for aromatic volatile organic compounds," *Sensors Actuators, B Chem.*, vol. 241, pp. 607–613, 2017, doi: 10.1016/j.snb.2016.10.124.

[18] Ruiyi, L., L. Ling, B. Hongxia, and L. Zaijun, "Nitrogen-doped multiple graphene aerogel/gold nanostar as the electrochemical sensing platform for ultrasensitive detection of circulating free DNA in human serum," *Biosens. Bioelectron.*, vol. 79, pp. 457–466, 2016, doi: 10.1016/j.bios.2015.12.092.

[19] Tian, Q., et al., "Three-dimensional PEDOT composite based electrochemical sensor for sensitive detection of chlorophenol," *J. Electroanal. Chem.*, vol. 837, no. November 2018, pp. 1–9, 2019, doi: 10.1016/j.jelechem.2019.01.055.

[20] Yang, J., et al., "Versatile Aerogels for Sensors," *Small*, vol. 15, no. 41, 2019, doi: 10.1002/smll.201902826.

[21] Li, Y. K., D. K. Yang, Y. C. Chen, H. J. Su, J. C. Wu, and Y. W. Chen-Yang, "A novel three-dimensional aerogel biochip for molecular recognition of nucleotide acids," *Acta Biomater.*, vol. 6, no. 4, pp. 1462–1470, 2010, doi: 10.1016/j.actbio.2009.10.001.

[22] Fang, L. X., K. J. Huang, and Y. Liu, "Novel electrochemical dual-aptamer-based sandwich biosensor using molybdenum disulfide/carbon aerogel composites and Au nanoparticles for signal amplification," *Biosens. Bioelectron.*, vol. 71, pp. 171–178, 2015, doi: 10.1016/j.bios.2015.04.031.

[23] Fort, C. I., R. Ortiz, L. C. Cotet, V. Danciu, I. C. Popescu, and L. Gorton, "Carbon Aerogel as Electrode Material for Improved Direct Electron Transfer in Biosensors Incorporating Cellobiose Dehydrogenase," *Electroanalysis*, vol. 28, no. 10, pp. 2311–2319, 2016, doi: 10.1002/elan.201600219.

[24] Dong, S., N. Li, G. Suo, and T. Huang, "Inorganic/organic doped carbon aerogels as biosensing materials for the detection of hydrogen peroxide," *Anal. Chem.*, vol. 85, no. 24, pp. 11739–11746, 2013, doi: 10.1021/ac4015098.

[25] Thubsuang, U., D. Sukanan, S. Sahasithiwat, S. Wongkasemjit, and T. Chaisuwan, "Highly sensitive room temperature organic vapor sensor based on polybenzoxazine-derived carbon aerogel thin film composite," *Mater. Sci. Eng. B Solid-State Mater. Adv. Technol.*, vol. 200, pp. 67–77, 2015, doi: 10.1016/j.mseb.2015.06.010.

[26] Jia, H., Q. Tian, J. Xu, L. Lu, X. Ma, and Y. Yu, "Aerogels prepared from polymeric β-cyclodextrin and graphene aerogels as a novel host-guest system for immobilization of antibodies: a voltammetric immunosensor for the tumor marker CA 15–3," *Microchim. Acta*, vol. 185, no. 11, pp. 1–7, 2018, doi: 10.1007/s00604-018-3056-3.

[27] Kaur, A., J. Kaur, and R. C. Singh, "Graphene aerogel based room temperature chemiresistive detection of hydrogen peroxide: A key

explosive ingredient," *Sensors Actuators, A Phys.*, vol. 282, pp. 97–113, 2018, doi: 10.1016/j.sna.2018.09.033.

[28] Tian, Y., et al., "Three-dimensional phosphorus-doped graphene as an efficient metal-free electrocatalyst for electrochemical sensing," *Sensors Actuators, B Chem.*, vol. 241, pp. 584–591, 2017, doi: 10.1016/j.snb.2016.10.113.

[29] Turner, S., et al., "Boron Doping and Defect Engineering of Graphene Aerogels for Ultrasensitive NO2 Detection," *J. Phys. Chem. C*, vol. 122, no. 35, pp. 20358–20365, 2018, doi: 10.1021/acs.jpcc.8b05984.

[30] Soorbaghi, F. P., M. Isanejad, S. Salatin, M. Ghorbani, S. Jafari, and H. Derakhshankhah, "Bioaerogels: Synthesis approaches, cellular uptake, and the biomedical applications," *Biomed. Pharmacother.*, vol. 111, no. January, pp. 964–975, 2019, doi: 10.1016/j.biopha.2019.01.014.

[31] Edwards, J. V., K. Fontenot, F. Liebner, N. D. Nee Pircher, A. D. French, and B. D. Condon, "Structure/function analysis of cotton-based peptide-cellulose conjugates: Spatiotemporal/kinetic assessment of protease aerogels compared to nanocrystalline and paper cellulose," *Int. J. Mol. Sci.*, vol. 19, no. 3, pp. 1–16, 2018, doi: 10.3390/ijms19030840.

[32] Wang, Y., Y. Su, W. Wang, Y. Fang, S. B. Riffat, and F. Jiang, "The advances of polysaccharide-based aerogels: Preparation and potential application," *Carbohydr. Polym.*, vol. 226, no. April, p. 115242, 2019, doi: 10.1016/j.carbpol.2019.115242.

[33] Esquivel-Castro, T. A., M. C. Ibarra-Alonso, J. Oliva, and A. Martínez-Luévanos, "Porous aerogel and core/shell nanoparticles for controlled drug delivery: A review," *Mater. Sci. Eng. C*, vol. 96, pp. 915–940, 2019, doi: 10.1016/j.msec.2018.11.067.

[34] Karamikamkar, S., H. E. Naguib, and C. B. Park, "Advances in precursor system for silica-based aerogel production toward improved mechanical properties, customized morphology, and multifunctionality: A review," *Adv. Colloid Interface Sci.*, vol. 276, p. 102101, 2020, doi: 10.1016/j.cis.2020.102101.

[35] Ali, I., et al., "Humidity-Responsive Gold Aerogel for Real-Time Monitoring of Human Breath," *Langmuir*, vol. 34, no. 16, pp. 4908–4913, 2018, doi: 10.1021/acs.langmuir.8b00472.

[36] Hosseini, H., M. Kokabi, and S. M. Mousavi, *BC/rGO conductive nanocomposite aerogel as a strain sensor*, vol. 137. Elsevier Ltd, 2018.

[37] Cho, H. J., I. D. Kim, and S. M. Jung, "Multifunctional Inorganic Nanomaterial Aerogel Assembled into fSWNT Hydrogel Platform for Ultraselective NO2 Sensing," *ACS Appl. Mater. Interfaces*, vol. 12, no. 9, pp. 10637–10647, 2020, doi: 10.1021/acsami.9b21174.

[38] Ding, A., B. Wang, J. Zheng, B. Weng, and C. Li, "Sensitive dopamine sensor based on three dimensional and macroporous carbon aerogel microelectrode," *Int. J. Electrochem. Sci.*, vol. 13, no. 5, pp. 4379–4389, 2018, doi: 10.20964/2018.05.43.

[39] Ai, S., et al., "Facile synthesis of nitrogen-doped graphene aerogels for electrochemical detection of dopamine," *Solid State Sci.*, vol. 86, no. October, pp. 6–11, 2018, doi: 10.1016/j.solidstatesciences.2018.09.014.

[40] Wang, M., Y. Chen, Y. Qin, T. Wang, J. Yang, and F. Xu, "Compressible, fatigue resistant, and pressure-sensitive carbon aerogels developed with a facile method for sensors and electrodes," *ACS Sustain. Chem. Eng.*, vol. 7, no. 15, pp. 12726–12733, 2019, doi: 10.1021/acssuschemeng.9b00814.

[41] Saadatnia, Z., S. G. Mosanenzadeh, T. Li, E. Esmailzadeh, and H. E. Naguib, "Polyurethane aerogel-based triboelectric nanogenerator for high performance energy harvesting and biomechanical sensing," *Nano Energy*, vol. 65, no. July, p. 104019, 2019, doi: 10.1016/j.nanoen.2019.104019.

[42] Krittayavathananon, A., and M. Sawangphruk, "Impedimetric Sensor of ss-HSDNA/Reduced Graphene Oxide Aerogel Electrode toward Aflatoxin B1 Detection: Effects of Redox Mediator Charges and Hydrodynamic Diffusion," *Anal. Chem.*, vol. 89, no. 24, pp. 13283–13289, 2017, doi: 10.1021/acs.analchem.7b03329.

[43] Martín-Pacheco, A., et al., "Graphene Quantum Dot-Aerogel: From Nanoscopic to Macroscopic Fluorescent Materials. Sensing Polyaromatic Compounds in Water," *ACS Appl. Mater. Interfaces*, vol. 10, no. 21, pp. 18192–18201, 2018, doi: 10.1021/acsami.8b02162.

[44] Ruiyi, L., et al., "Novel graphene-gold nanohybrid with excellent electrocatalytic performance for the electrochemical detection of glucose," *Sensors Actuators, B Chem.*, vol. 208, pp. 421–428, 2015, doi: 10.1016/j.snb.2014.11.004.

[45] Yan, W., et al., "Conductometric gas sensing behavior of WS2 aerogel," *FlatChem*, vol. 5, pp. 1–8, 2017, doi: 10.1016/j.flatc.2017.08.003.

[46] Liu, L., et al., "One-pot synthesis of 3-dimensional reduced graphene oxide-based hydrogel as support for microbe immobilization and BOD biosensor preparation," *Biosens. Bioelectron.*, vol. 63, pp. 483–489, 2015, doi: 10.1016/j.bios.2014.07.074.

[47] Salek-Maghsoudi, A., et al., "Recent advances in biosensor technology in assessment of early diabetes biomarkers," *Biosens. Bioelectron.*, vol. 99, pp. 122–135, 2018, doi: 10.1016/j.bios.2017.07.047.

[48] Sabu, C., T. K. Henna, V. R. Raphey, K. P. Nivitha, and K. Pramod, "Advanced biosensors for glucose and insulin," *Biosens. Bioelectron.*, vol. 141, no. January, p. 111201, 2019, doi: 10.1016/j.bios.2019.03.034.

[49] Gao, Y., et al., "Three-dimensional porous Cu@Cu 2 O aerogels for direct voltammetric sensing of glucose," *Microchim. Acta*, vol. 186, no. 3, pp. 1–9, 2019, doi: 10.1007/s00604-019-3263-6.

[50] Ling, P., Q. Zhang, T. Cao, and F. Gao, "Versatile Three-Dimensional Porous Cu@Cu 2 O Aerogel Networks as Electrocatalysts and Mimicking Peroxidases," *Angew. Chemie*, vol. 130, no. 23, pp. 6935–6940, 2018, doi: 10.1002/ange.201801369.

[51] Wang, B., et al., "Fabrication of graphene aerogel/platinum nanoparticle hybrids for the direct electrochemical analysis of glucose," *J. Nanosci. Nanotechnol.*, vol. 16, no. 7, pp. 6895–6902, 2016, doi: 10.1166/jnn.2016.11359.

[52] Wang, B., S. Yan, and Y. Shi, "Direct electrochemical analysis of glucose oxidase on a graphene aerogel/gold nanoparticle hybrid for glucose biosensing," *J. Solid State Electrochem.*, vol. 19, no. 1, pp. 307–314, 2014, doi: 10.1007/s10008-014-2608-7.

[53] Ma, C. B., Y. Zhang, Q. Liu, Y. Du, and E. Wang, "Enhanced Stability of Enzyme Immobilized in Rationally Designed Amphiphilic Aerogel and Its Application for Sensitive Glucose Detection," *Anal. Chem.*, 2020, doi: 10.1021/acs.analchem.9b05858.

[54] Li, R., T. Yang, Z. Li, Z. Gu, G. Wang, and J. Liu, "Synthesis of palladium@gold nanoalloys/nitrogen and sulphur-functionalized multiple graphene aerogel for electrochemical detection of dopamine," *Anal. Chim. Acta*, vol. 954, pp. 43–51, 2017, doi: 10.1016/j.aca.2016.12.031.

[55] Ruiyi, L., Q. Sili, L. Zhangyi, L. Ling, and L. Zaijun, "Histidine-functionalized graphene quantum dot-graphene micro-aerogel based voltammetric sensing of dopamine," *Sensors Actuators, B Chem.*, vol. 250, pp. 372–382, 2017, doi: 10.1016/j.snb.2017.05.001.

[56] Yang, Z., X. Liu, Y. Wu, and C. Zhang, "Modification of carbon aerogel electrode with molecularly imprinted polypyrrole for electrochemical determination of dopamine," *Sensors Actuators, B Chem.*, vol. 212, pp. 457–463, 2015, doi: 10.1016/j.snb.2015.02.057.

[57] Li, Y., et al., "Flexible chitosan/carbon nanotubes aerogel, a robust matrix for in-situ growth and non-enzymatic biosensing applications," *Sensors Actuators, B Chem.*, vol. 232, pp. 750–757, 2016, doi: 10.1016/j.snb.2016.04.023.

[58] Chen, M., C. Hou, D. Huo, H. Fa, Y. Zhao, and C. Shen, "A sensitive electrochemical DNA biosensor based on three-dimensional nitrogen-doped graphene and Fe3O4 nanoparticles," *Sensors Actuators, B Chem.*, vol. 239, pp. 421–429, 2017, doi: 10.1016/j.snb.2016.08.036.

[59] Ruiyi, L., W. Jiajia, L. Ling, and L. Zaijun, "Ultrasensitive direct detection of dsDNA using a glassy carbon electrode modified with thionin-

functionalized multiple graphene aerogel and gold nanostars," *Microchim. Acta*, vol. 183, no. 5, pp. 1641–1649, 2016, doi: 10.1007/s00604-016-1793-8.

[60] Yuanfeng, P., L. Ruiyi, S. Xiulan, W. Guangli, and L. Zaijun, "Highly sensitive electrochemical detection of circulating tumor DNA in human blood based on urchin-like gold nanocrystal-multiple graphene aerogel and target DNA-induced recycling double amplification strategy," *Anal. Chim. Acta*, vol. 1121, pp. 17–25, 2020, doi: 10.1016/j.aca.2020.04.077.

[61] Jia, H., et al., "Three-dimensional Au nanoparticles/nano-poly(3,4-ethylene dioxythiophene)- graphene aerogel nanocomposite: A high-performance electrochemical immunosensing platform for prostate specific antigen detection," *Sensors Actuators, B Chem.*, vol. 260, pp. 990–997, 2018, doi: 10.1016/j.snb.2018.01.006.

[62] Xie, B., H. Peng, C. Wang, Z. Zhang, and Y. He, "Controlled Coassembly of Dumbbell-Like Au Nanoparticles with a Porous Nitrogen Doped Carbon Aerogel for Cancer Cell H2O2 Detection," *Anal. Chim. Acta*, 2020, doi: 10.1016/j.aca.2020.06.003.

[63] Zhao, Y., et al., "Electrochemical biointerface based on electrodeposition AuNPs on 3D graphene aerogel: Direct electron transfer of Cytochrome c and hydrogen peroxide sensing," *J. Electroanal. Chem.*, vol. 842, no. December 2018, pp. 16–23, 2019, doi: 10.1016/j.jelechem.2019.04.052.

[64] Yan, S., et al., "Enhanced nonenzymatic sensing of hydrogen peroxide released from living cells based on graphene aerogel/platinum nanoparticle," *Sci. Adv. Mater.*, vol. 8, no. 6, pp. 1165–1171, 2016, doi: 10.1166/sam.2016.2701.

[65] Baia, M., V. Danciu, V. Cosoveanu, and L. Baia, "Porous nanoarchitectures based on TiO2 aerogels and Au particles as potential SERS sensor for monitoring of water quality," *Vib. Spectrosc.*, vol. 48, no. 2, pp. 206–209, 2008, doi: 10.1016/j.vibspec.2008.03.005.

[66] Liu, Y., and M. Wei, "Development of acetylcholinesterase biosensor based on platinum-carbon aerogels composite for determination of organophosphorus pesticides," *Food Control*, vol. 36, no. 1, pp. 49–54, 2014, doi: 10.1016/j.foodcont.2013.08.005.

[67] Kehrle, J., et al., "Superhydrophobic Silicon Nanocrystal-Silica Aerogel Hybrid Materials: Synthesis, Properties, and Sensing Application," *Langmuir*, vol. 34, no. 16, pp. 4888–4896, 2018, doi: 10.1021/acs.langmuir.7b03746.

[68] Hongxia, C., H. Ji, L. Zaijun, L. Ruiyi, Y. Yongqiang, and S. Xiulan, "Electrochemical aptasensor for detection of acetamiprid in vegetables with graphene aerogel-glutamic acid functionalized graphene quantum

dot/gold nanostars as redox probe with catalyst," *Sensors Actuators, B Chem.*, vol. 298, no. May, 2019, doi: 10.1016/j.snb.2019.126866.

[69] Hu, T., et al., "Visual detection of mixed organophosphorous pesticide using QD-AChE aerogel based microfluidic arrays sensor," *Biosens. Bioelectron.*, vol. 136, no. January, pp. 112–117, 2019, doi: 10.1016/j.bios.2019.04.036.

[70] Xie, Y., et al., "In-situ synthesis of hierarchically porous polypyrrole@ZIF-8/graphene aerogels for enhanced electrochemical sensing of 2, 2-methylenebis (4-chlorophenol)," *Electrochim. Acta*, vol. 311, pp. 114–122, 2019, doi: 10.1016/j.electacta.2019.04.132.

[71] Zhang, Y., et al., "In Situ Growth of Fe2O3 Nanoparticles on Highly Porous Graphene/Polyimide-Based Carbon Aerogel Nanocomposites for Effectively Selective Detection of Dopamine," *Adv. Mater. Interfaces*, vol. 3, no. 15, 2016, doi: 10.1002/admi.201600137.

[72] Ruiyi, L., Z. Haiyan, L. Zaijun, and L. Junkang, "Electrochemical determination of acetaminophen using a glassy carbon electrode modified with a hybrid material consisting of graphene aerogel and octadecylamine-functionalized carbon quantum dots," *Microchim. Acta*, vol. 185, no. 2, pp. 1–9, 2018, doi: 10.1007/s00604-018-2688-7.

[73] Sun, Y., et al., "Highly selective and sensitive streptomycin chemiluminescence sensor based on aptamer and G-quadruplex DNAzyme modified three-dimensional graphene composite," *Sensors Actuators, B Chem.*, vol. 301, no. September, p. 127122, 2019, doi: 10.1016/j.snb.2019.127122.

[74] Yin, D., X. Bo, J. Liu, and L. Guo, "A novel enzyme-free glucose and H2O2 sensor based on 3D graphene aerogels decorated with Ni3N nanoparticles," *Anal. Chim. Acta*, vol. 1038, pp. 11–20, 2018, doi: 10.1016/j.aca.2018.06.086.

[75] Sun, Q. Q., M. Xu, S. J. Bao, and C. Ming Li, "PH-controllable synthesis of unique nanostructured tungsten oxide aerogel and its sensitive glucose biosensor," *Nanotechnology*, vol. 26, no. 11, p. 115602, 2015, doi: 10.1088/0957-4484/26/11/115602.

[76] Xu, J., et al., "3D Porous Graphene Aerogel@GOx Based Microfluidic Biosensor for Electrochemical Glucose Detection," *Analyst*, 2020, doi: 10.1039/d0an00681e.

[77] Jeong, J. M., M. H. Yang, D. S. Kim, T. J. Lee, B. G. Choi, and D. H. Kim, "High performance electrochemical glucose sensor based on three-dimensional MoS 2/graphene aerogel," *J. Colloid Interface Sci.*, vol. 506, pp. 379–385, 2017, doi: 10.1016/j.jcis.2017.07.061.

[78] Zhang, C., S. Si, and Z. Yang, "Design of molecularly imprinted TiO2/carbon aerogel electrode for the photoelectrochemical determination

of atrazine," *Sensors Actuators, B Chem.*, vol. 211, pp. 206–212, 2015, doi: 10.1016/j.snb.2015.01.079.

[79] Ruiyi, L., C. Fangchao, Z. Haiyan, S. Xiulan, and L. Zaijun, "Electrochemical sensor for detection of cancer cell based on folic acid and octadecylamine-functionalized graphene aerogel microspheres," *Biosens. Bioelectron.*, vol. 119, pp. 156–162, 2018, doi: 10.1016/j.bios.2018.07.060.

[80] Wu, Y., et al., "Polydopamine-Capped Bimetallic AuPt Hydrogels Enable Robust Biosensor for Organophosphorus Pesticide Detection," *Small*, vol. 15, no. 17, pp. 1–8, 2019, doi: 10.1002/smll.201900632.

[81] Liu, D., Y. Wang, and G. Zhao, "Preparation of graphene aerogel for determining oxalic acid," *Int. J. Electrochem. Sci.*, vol. 10, no. 8, pp. 6794–6802, 2015.

[82] Li, R., L. Liu, H. Zhu, and Z. Li, "Synthesis of gold-palladium nanowaxberries/dodecylamine-functionalized graphene quantum dots-graphene micro-aerogel for voltammetric determination of peanut allergen Ara h 1," *Anal. Chim. Acta*, vol. 1008, pp. 38–47, 2018, doi: 10.1016/j.aca.2018.01.031.

[83] Yang, L., et al., "3D Nanostructured Palladium-Functionalized Graphene-Aerogel-Supported Fe3O4 for Enhanced Ru(bpy)32+-Based Electrochemiluminescent Immunosensing of Prostate Specific Antigen," *ACS Appl. Mater. Interfaces*, vol. 9, no. 40, pp. 35260–35267, 2017, doi: 10.1021/acsami.7b11458.

[84] Edwards, J. V., K. R. Fontenot, F. Liebner, and B. D. Condon, "Peptide-cellulose conjugates on cotton-based materials have protease sensor/sequestrant activity," *Sensors (Switzerland)*, vol. 18, no. 7, pp. 1–16, 2018, doi: 10.3390/s18072334.

[85] Edwards, V. J., K. R. Fontenot, N. T. Prevost, N. Pircher, F. Liebner, and B. D. Condon, "Preparation, characterization and activity of a peptide-cellulosic aerogel protease sensor from cotton," *Sensors (Switzerland)*, vol. 16, no. 11, pp. 1–19, 2016, doi: 10.3390/s16111789.

[86] Wang, B., S. Yan, Z. Lin, Y. Shi, D. Hu, and X. Xu, "Direct electrochemistry of cytochrome c on graphene aerogel/gold nanoparticle hybrid for hydrogen peroxide biosensing," *J. Bionanoscience*, vol. 9, no. 5, pp. 330–334, 2015, doi: 10.1166/jbns.2015.1318.

[87] Du, X., et al., "Building a Three-Dimensional Nano-Bio Interface for Aptasensing: An Analytical Methodology Based on Steric Hindrance Initiated Signal Amplification Effect," *Anal. Chem.*, vol. 88, no. 19, pp. 9622–9629, 2016, doi: 10.1021/acs.analchem.6b02368.

[88] Tian, J., D. Wang, Y. Zheng, and T. Jing, "A high sensitive electrochemical avian influenza virus H7 biosensor based on CNTs/MoSx aerogel," *Int. J.*

Electrochem. Sci., vol. 12, no. 4, pp. 2658–2668, 2017, doi: 10.20964/2017.04.30.

[89] He, X., et al., "Strategy of Constructing Light-Weight and Highly Compressible Graphene-Based Aerogels with an Ordered Unique Configuration for Wearable Piezoresistive Sensors," *ACS Appl. Mater. Interfaces*, vol. 11, no. 21, pp. 19350–19362, 2019, doi: 10.1021/acsami.9b02591.

[90] Chen, X., et al., "Highly Compressible and Robust Polyimide/Carbon Nanotube Composite Aerogel for High-Performance Wearable Pressure Sensor," *ACS Appl. Mater. Interfaces*, vol. 11, no. 45, pp. 42594–42606, 2019, doi: 10.1021/acsami.9b14688.

[91] Chen, Z., et al., "Wood-Derived Lightweight and Elastic Carbon Aerogel for Pressure Sensing and Energy Storage," *Adv. Funct. Mater.*, vol. 30, no. 17, pp. 1–11, 2020, doi: 10.1002/adfm.201910292.

[92] Qian, C., et al., "All-printed 3D hierarchically structured cellulose aerogel based triboelectric nanogenerator for multi-functional sensors," *Nano Energy*, vol. 63, no. June, p. 103885, 2019, doi: 10.1016/j.nanoen.2019.103885.

[93] Huang, J., Z. Li, X. Wu, J. Wang, and S. Yang, "Poly(vinyl alcohol)-Mediated Graphene Aerogels with Tailorable Architectures and Advanced Properties for Anisotropic Sensing," *J. Phys. Chem. C*, vol. 123, no. 6, pp. 3781–3789, 2019, doi: 10.1021/acs.jpcc.8b11327.

[94] Qian, Z., et al., "Triboelectric nanogenerators made of polybenzazole aerogels as fire-resistant negative tribo-materials," *Nano Energy*, vol. 64, no. June, p. 103900, 2019, doi: 10.1016/j.nanoen.2019.103900.

[95] Huang, J., et al., "Multi-Arch-Structured All-Carbon Aerogels with Superelasticity and High Fatigue Resistance as Wearable Sensors," *ACS Appl. Mater. Interfaces*, vol. 12, no. 14, pp. 16822–16830, 2020, doi: 10.1021/acsami.0c01794.

[96] Long, S., et al., "An ultralight, supercompressible, superhydrophobic and multifunctional carbon aerogel with a specially designed structure," *Carbon N. Y.*, vol. 158, pp. 137–145, 2020, doi: 10.1016/j.carbon.2019.11.065.

[97] Luo, Q., et al., "Carbon Nanotube/Chitosan-Based Elastic Carbon Aerogel for Pressure Sensing," *Ind. Eng. Chem. Res.*, vol. 58, no. 38, pp. 17768–17775, 2019, doi: 10.1021/acs.iecr.9b02847.

[98] Ma, Y., et al., "3D Synergistical MXene/Reduced Graphene Oxide Aerogel for a Piezoresistive Sensor," *ACS Nano*, vol. 12, no. 4, pp. 3209–3216, 2018, doi: 10.1021/acsnano.7b06909.

[99] Xu, X., et al., "Copper Nanowire-Based Aerogel with Tunable Pore Structure and Its Application as Flexible Pressure Sensor," *ACS Appl.*

Mater. Interfaces, vol. 9, no. 16, pp. 14273–14280, 2017, doi: 10.1021/acsami.7b02087.

[100] Yap, L. W., S. Gong, Y. Tang, Y. Zhu, and W. Cheng, "Soft piezoresistive pressure sensing matrix from copper nanowires composite aerogel," *Sci. Bull.*, vol. 61, no. 20, pp. 1624–1630, 2016, doi: 10.1007/s11434-016-1149-0.

[101] Zhao, S., et al., "Strain-driven and ultrasensitive resistive sensor/switch based on conductive alginate/nitrogen-doped carbon-nanotube-supported Ag hybrid aerogels with pyramid design," *ACS Appl. Mater. Interfaces*, vol. 6, no. 24, pp. 22823–22829, 2014, doi: 10.1021/am5069936.

[102] Zhu, M., et al., "Hollow MXene Sphere/Reduced Graphene Aerogel Composites for Piezoresistive Sensor with Ultra-High Sensitivity," *Adv. Electron. Mater.*, vol. 6, no. 2, pp. 1–9, 2020, doi: 10.1002/aelm.201901064.

[103] Zu, G., K. Kanamori, K. Nakanishi, and J. Huang, "Superhydrophobic Ultraflexible Triple-Network Graphene/Polyorganosiloxane Aerogels for a High-Performance Multifunctional Temperature/Strain/Pressure Sensing Array," *Chem. Mater.*, vol. 31, no. 16, pp. 6276–6285, 2019, doi: 10.1021/acs.chemmater.9b02437.

[104] Huang, P., et al., "Dual-Mode Carbon Aerogel/Iron Rubber Sensor," *ACS Appl. Mater. Interfaces*, vol. 12, no. 7, pp. 8674–8680, 2020, doi: 10.1021/acsami.9b20662.

[105] Chen, Z., et al., "Compressible, Elastic, and Pressure-Sensitive Carbon Aerogels Derived from 2D Titanium Carbide Nanosheets and Bacterial Cellulose for Wearable Sensors," *Chem. Mater.*, vol. 31, no. 9, pp. 3301–3312, 2019, doi: 10.1021/acs.chemmater.9b00259.

[106] Huang, J., et al., "Flexible electrically conductive biomass-based aerogels for piezoresistive pressure/strain sensors," *Chem. Eng. J.*, vol. 373, no. May, pp. 1357–1366, 2019, doi: 10.1016/j.cej.2019.05.136.

[107] Hosseini, H., M. Kokabi, and S. M. Mousavi, "Conductive bacterial cellulose/multiwall carbon nanotubes nanocomposite aerogel as a potentially flexible lightweight strain sensor," *Carbohydr. Polym.*, vol. 201, no. August, pp. 228–235, 2018, doi: 10.1016/j.carbpol.2018.08.054.

[108] Li, W., et al., "Flexible strain sensor based on aerogel-spun carbon nanotube yarn with a core-sheath structure," *Compos. Part A Appl. Sci. Manuf.*, vol. 108, pp. 107–113, 2018, doi: 10.1016/j.compositesa.2018.02.024.

[109] Wu, S., et al., "Strain Sensors with Adjustable Sensitivity by Tailoring the Microstructure of Graphene Aerogel/PDMS Nanocomposites," *ACS Appl. Mater. Interfaces*, vol. 8, no. 37, pp. 24853–24861, 2016, doi: 10.1021/acsami.6b06012.

[110] Zhang, H., et al., "Metallic Sandwiched-Aerogel Hybrids Enabling Flexible and Stretchable Intelligent Sensor," *Nano Lett.*, 2020, doi: 10.1021/acs.nanolett.0c00372.

[111] Zhang, F., et al., "Multi-modal strain and temperature sensor by hybridizing reduced graphene oxide and PEDOT:PSS," *Compos. Sci. Technol.*, vol. 187, no. September 2019, p. 107959, 2020, doi: 10.1016/j.compscitech.2019.107959.

[112] Hou, X., R. Zhang, and D. Fang, "Superelastic, fatigue resistant and heat insulated carbon nanofiber aerogels for piezoresistive stress sensors," *Ceram. Int.*, vol. 46, no. 2, pp. 2122–2127, 2020, doi: 10.1016/j.ceramint.2019.09.195.

[113] Wei, N., et al., "Compressible Supercapacitor with Residual Stress Effect for Sensitive Elastic-Electrochemical Stress Sensor," *ACS Appl. Mater. Interfaces*, vol. 10, no. 44, pp. 38057–38065, 2018, doi: 10.1021/acsami.8b12745.

[114] Wang, C. T., C. L. Wu, I. C. Chen, and Y. H. Huang, "Humidity sensors based on silica nanoparticle aerogel thin films," *Sensors Actuators, B Chem.*, vol. 107, no. 1 SPEC. ISS., pp. 402–410, 2005, doi: 10.1016/j.snb.2004.10.034.

[115] Ayers, M. R., and A. J. Hunt, "Molecular oxygen sensors based on photoluminescent silica aerogels," *J. Non. Cryst. Solids*, vol. 225, no. 1–3, pp. 343–347, 1998, doi: 10.1016/S0022-3093(98)00051-9.

[116] Ma, Y., X. Chen, J. Bai, G. Yuan, and L. Ren, "Highly selective fluorescence chemosensor based on carbon-dot-aerogel for detection of aniline gas," *Inorg. Chem. Commun.*, vol. 100, no. December 2018, pp. 64–69, 2019, doi: 10.1016/j.inoche.2018.12.013.

[117] Chung, V. P. J., M. C. Yip, and W. Fang, "Resorcinol-formaldehyde aerogels for CMOS-MEMS capacitive humidity sensor," *Sensors Actuators, B Chem.*, vol. 214, pp. 181–188, 2015, doi: 10.1016/j.snb.2015.03.024.

[118] Wang, R., G. Li, Y. Dong, Y. Chi, and G. Chen, "Carbon quantum dot-functionalized aerogels for NO2 gas sensing," *Anal. Chem.*, vol. 85, no. 17, pp. 8065–8069, 2013, doi: 10.1021/ac401880h.

[119] Long, H., et al., "High Surface Area MoS2/Graphene Hybrid Aerogel for Ultrasensitive NO2 Detection," *Adv. Funct. Mater.*, vol. 26, no. 28, pp. 5158–5165, 2016, doi: 10.1002/adfm.201601562.

[120] Long, H., et al., "3D MoS2 Aerogel for Ultrasensitive NO2 Detection and Its Tunable Sensing Behavior," *Adv. Mater. Interfaces*, vol. 4, no. 16, pp. 1–8, 2017, doi: 10.1002/admi.201700217.

[121] Liu, X., J. Cui, J. Sun, and X. Zhang, "3D graphene aerogel-supported SnO2 nanoparticles for efficient detection of NO2," *RSC Adv.*, vol. 4, no. 43, pp. 22601–22605, 2014, doi: 10.1039/c4ra02453b.
[122] Li, L., S. He, M. Liu, C. Zhang, and W. Chen, "Three-dimensional mesoporous graphene aerogel-supported SnO2 nanocrystals for high-performance NO2 gas sensing at low temperature," *Anal. Chem.*, vol. 87, no. 3, pp. 1638–1645, 2015, doi: 10.1021/ac503234e.
[123] Li, Q., et al., "Ag modified 3D reduced graphene oxide aerogel based sensor with embedded microheater for fast response and high-sensitive detection of NO 2," 2020, doi: 10.1021/acsami.9b22098.
[124] Yan, W., M. A. Worsley, T. Pham, A. Zettl, C. Carraro, and R. Maboudian, "Effects of ambient humidity and temperature on the NO 2 sensing characteristics of WS 2/graphene aerogel," *Appl. Surf. Sci.*, vol. 450, pp. 372–379, 2018, doi: 10.1016/j.apsusc.2018.04.185.
[125] Shu, J., Z. Qiu, and D. Tang, "Self-Referenced Smartphone Imaging for Visual Screening of H2S Using CuxO-Polypyrrole Conductive Aerogel Doped with Graphene Oxide Framework," *Anal. Chem.*, vol. 90, no. 16, pp. 9691–9694, 2018, doi: 10.1021/acs.analchem.8b03011.
[126] Alizadeh, T., and F. Ahmadian, "Thiourea-treated graphene aerogel as a highly selective gas sensor for sensing of trace level of ammonia," *Anal. Chim. Acta*, vol. 897, pp. 87–95, 2015, doi: 10.1016/j.aca.2015.09.031.
[127] Ma, X., et al., "Sensitive determination of nitrite by using an electrode modified with hierarchical three-dimensional tungsten disulfide and reduced graphene oxide aerogel," *Microchim. Acta*, vol. 186, no. 5, pp. 1–9, 2019, doi: 10.1007/s00604-019-3379-8.
[128] Li, Y. K., Y. C. Chen, K. J. Jiang, J. chuang Wu, and Y. W. Chen-Yang, "Three-dimensional arrayed amino aerogel biochips for molecular recognition of antigens," *Biomaterials*, vol. 32, no. 30, pp. 7347–7354, 2011, doi: 10.1016/j.biomaterials.2011.06.069.
[129] You, X., J. Wu, and Y. Chi, "Superhydrophobic Silica Aerogels Encapsulated Fluorescent Perovskite Quantum Dots for Reversible Sensing of SO 2 in a 3D-Printed Gas Cell," *Anal. Chem.*, vol. 91, no. 8, pp. 5058–5066, 2019, doi: 10.1021/acs.analchem.8b05253.

Chapter 7

Synthesis and Applications of Nanocarbon Aerogels

Jair A. Yánez Sánchez[1], Nelly Ma. Rosas-Laverde[1] and Alina I. Pruna[2,3,*]

[1]Department of Materials, Escuela Politécnica Nacional, Quito, Ecuador
[2]Institute of Materials Technology, Universitat Politecnica de Valencia, Valencia, Spain
[3]Center for Surface Science and Nanotechnology, University Politehnica of Bucharest, Bucharest, Romania

Abstract

With the advance in the development of material synthesis, it has been possible to integrate the use of the different allotropic forms of carbon in the development of aerogels, thus combining the excellent individual properties of these materials within an aerogel form. Aerogels exhibit interesting properties including high porosity, high surface area, low density that make them an excellent alternative for many fields such as air and water pollutant removal, thermal insulation or energy storage. This chapter highlights the synthesis methods of aerogels obtained from the main nanocarbon materials including carbon nanotubes, graphene and nanodiamonds and the research results obtained by application of such aerogels in varying fields.

Keywords: aerogel, carbon nanotubes, graphene, nanodiamonds, synthesis, applications

---

* Corresponding Author's Email: apruna@itm.upv.es.

In: Aerogels: Properties and Applications
Editor: Alina Iuliana Pruna
ISBN: 978-1-68507-788-4

Introduction

Aerogels are synthetic materials with high porosity, which gives them the main characteristic of being ultra-light materials. These are obtained by gelation processes from a wide variety of materials, in which the suspended phase forms interlocking structures [1], followed by a subsequent process of replacement of the liquid phase of the gel with another substance to be further evaporated, without damaging the network. The material that is obtained has a large amount of mesopores and micropores [2], air accounting for about 60% to 90% of the volume, hence its name. Carbon aerogels represent a subgroup of aerogels characterized in that their constituent matrix is made from a form or combination of varying allotropic forms of carbon. These aerogels are characterised by a high surface area, which in combination with the nonpolar characteristic of carbon, make it ideal for the use in hydrophobic applications [3]. This type of materials exhibits, beside the large specific surface area, also a high electrical conductivity, as well as being chemically inert, which is why they aroused research interest for various applications [4].

These materials were synthesized for the first time at the end of the 80's by Professor Pekala by controlled pyrolysis of polymeric aerogels. Pekala generally started from the sol-gel polycondensation of Resorcinol-Formaldehyde (RF) mixtures [5] for the production of these aerogels, to then be carbonized, thus leaving a nano-structured amorphous carbon matrix. Certain properties, such as the density and pore size can be controlled by varying the conditions in which the sol-gel process occurs [6].

The carbon-based aerogels show enormous potential in various applications, especially energy and environmental remediation [7], branches that have gained increased interest from the scientific community, as it is evident from the bibliographic review of the publications reated to these materials. As depicted in Figure 1, an exponential growth in scientific publications has been observed since 2010, motivated by the need to find solutions to problems related to the previously mentioned topics.

With the advances in material synthesis, it has been possible to integrate the use of different allotropic forms of carbon in the development of aerogels, thus combining the excellent individual properties of these materials with an aerogel form, and consequently to improve their applicability.

Nanoparticles and nanocomposites have been shown to contribute enormously with new characteristics that could not be obtained directly with bulk materials. By taking this aspect into account, the use of nanoparticles in macroscopic structures has been investigated as an alternative approach to improve physical properties that could be used for the development of materials with characteristics tailored to needs [8].

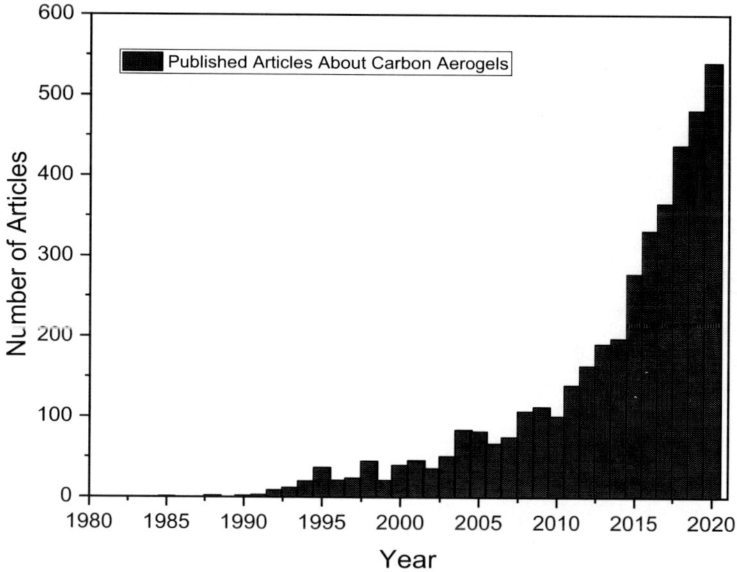

Figure 1. Number of publications made about carbon aerogels.

The main synthetic methods employed in the development of aerogels based on the main allotropic forms of carbon are addressed in the present chapter. Further, the properties exhibited by such aerogels are depicted, as well as the most relevant advances in the application fields that take advantage on these properties.

Graphene Aerogels: General Aspects and Synthesis

Graphene is an allotropic form of carbon; it can be described as a two-dimensional (2D) element which consists of a crystalline sheet of hexagonal lattice composed only of carbon atoms joined by covalent bonds. It is considered as an individual layer derived from a graphite structure [9]. Although carbon and other structures derived from it have been used for several decades, it was not until 2004 that the discovery of this type of material was achieved thanks to the work of Andre Geim and Konstantin Novoselov [10, 11]. The strong covalent bonds in the basal plane of graphene confer it with excellent properties, both mechanical and conductive, as it can be seen in the summary in Table 1. Its unique properties make it attractive for the study and use in applications such as energy storage [12–14], environmental remediation [15–18], catalysis [19–21], electronics [22–25], chemical applications [21, 26–30], biomedicine and biotechnology [16, 31–34].

Table 1. Main physical properties for single layer graphene

Property	Value	Reference
Young Modulus [Tpa]	1	[35]
Tensile Strength [Gpa]	130	[35]
Stiffness [Tpa]	0.27-1.47	[35]
Electrical Conductivity [MS/m]	100	[30]
Thermal Conductivity [W/mK]	3500 - 5000	[30]

Table 2. Main methods for graphene synthesis

Synthesis Method	Form	Size	Some highlights
Liquid Phase Exfoliation	Nanosheets	nm to a few μm	- High scalability - Low yield - Moderate quality - Low cost - Impurity
Chemical Reduction of Grafite Oxide	Nanoflakes Powder	nm to a few μm	- High scalability - Low cost - Low purity - High defect density
Chemical vapor deposition (on metalic Substrate)	Thin films	max. 75 cm	- Moderate scalability - High cost - High quality - High process temperature (>1000°C)
Unzipping CNT	Nanoribbons	few μm	- Moderate scalability - High yield - High quality - Potentially low cost
Epitaxial Growth (on SiC substrate)	Thin films	~ 50 μm	- Low yield - High quality - High cost - High process temperature (1500°C) - Expensive substrate
Micromechanical Exfoliation	Flakes	5 to 100 μm	- Small scale production - High cost - High quality - Uneven films

The synthesis method for this type of materials is of vital importance and it is directly related to the properties exhibited in varying applications. Some methods are shown in Table 2 [36]. The use of the 2D graphene sheets in a three-dimensional (3D) structure that can be replicated until obtaining a volumetric material is a great challenge that slowed down to some extent the potential of such material. However, as a solution approach to this problem, several options have been used, mainly incorporating it into polymers [37], where, unfortunately, it tends to pile up forming nano graphite sheets, significantly reducing its properties. For this reason, the study of the application of graphene as a constituent in aerogels has attracted great interest [1].

The synthetic techniques can be divided into direct and indirect methods of obtaining graphene aerogels, where the first group refers to the assembling the graphene nanomaterial directly until obtaining the volumetric material [38–40], whereas the indirect techniques are based on the use of precursors, which result in a three-dimensional structure of graphene nano-sheets by different chemical treatments [32, 33, 41, 42].

The template directed – chemical vapor deposition method, as its name implies, consists in using a template on which graphene is deposited by chemical vapour deposition [43]. For this purpose, a previously prepared porous material is used as a template, and a carbon-rich substance as a precursor. Chen et al., [44] reported on the production of porous graphene monoliths by using a nickel foam, in which it introduces methane at different concentrations at a temperature of 1000°C at atmospheric pressure. At such temperature, CH_4 decomposes into carbon, which is deposited on the surface of the template in the form of graphene nano-sheets. Similarly, Dong et al., [45] used ethanol as a precursor for the formation of graphene on nickel foams. Once the graphene was deposited on the template, a thin coating was applied to protect it from the necessary chemical attack to remove the template. Such layer is usually made of a thermoplastic such as polymethylmethacrylate. For the removal of the template, a chemical attack is used which makes use of HCl or $FeCl_3$ [44, 45]. Finally, the PMMA layer is removed, as well, with a hot solution of acetone.

Due to the deposition method and the difference between the coefficients of thermal expansion of graphene and the (nickel) foam, the surfaces of these monoliths present ripples and wrinkles, which give them a better surface condition in case of being used as a base for the application of a polymeric layer in case a composite is better performing [1].

The results of the chemical vapour deposition with the use of a template, shows a good quality graphene formation, which is indicated by an absence of the defects band (D peak by means of Raman spectroscopy measurement). Therefore, the graphene obtained by such method is considered of very high quality [45]. The specific surfaces measured for this material show a large surface area, where in the case of CH_4, where the layers are thinner, the surface area value is about 850 m^2g^{-1} [44], while for ethanol precursor, the surface area reaches 670 m^2g^{-1} [45].

In-situ reduction assembly method is a form of synthesis consisting in the preparation of a graphene hydrogel by adding the reducing agents in the graphene oxide precursor dispersion, thus, by hydrophobic interactions, and π-π stacking. The graphene nanosheets get assembled into three-dimensional structures within a hydrogel. Further, a freeze-drying technique is used, in such a way that the solvents are evaporated and only the graphene aerogel remains [1]. The preparation starts with the presence of the graphene oxide precursor, which is generally synthesized by the Hummers or Staudenmaier methods [46]. The graphene oxide is employed as a diluted aqueous dispersion, as reported by Zhang et al. [47] who used a dispersion of a concentration 0.1-4.5 mg mL^{-1}, mixing it with the reducing agent (oxalic acid in the presence of sodium iodide). The mixture was stirred by different methods such as mechanical and ultrasonic stirring, and then it was allowed to react at 90 ° C for 12 hours in an oil bath. Finally, a washing process was applied for the removal of impurities and the hydrogel was subjected to freeze-drying [47]. In this case, specific surfaces of 151 m^2g^{-1} have been achieved, which is very high for a porous material, although it is significantly lower than the material obtained by templated chemical vapour deposition, because the walls of the structure are thicker, being a product of a greater stacking of graphene sheets. The Raman spectroscopy measurements confirmed the mterial exhibited a pronounced D band; however, graphene could be formed by reduction. The reduction was also confirmed by Infrared Fourier Transform spectroscopy (FTIR) measurements, where a dramatic reduction of functional groups containing oxygen was observed. The compression tests indicated that the formed aerogel could withstand up to 7400 times its weight. Finally, the scanning electron microscopy (SEM) measurements showed a nanostructure from which the average size of the pores could be determined, which in the case of the use of oxalic acid and sodium iodide presented values of about 3-5 μm [47].

The use of different reducing agents is critical for the final properties of the aerogel [48]. Such agents include N_2H_4, $NaBH_4$, ammonia, vitamin C, etc.

They will influence the magnetic properties, surface characteristics, photocatalytic, specific surface properties, pore size, compressibility, etc. [49]. For example, Tang et al. [50] employed paraphenylene diamine (PPD) in the presence of 135 μL of ammonia, for a 5 mL solution of graphene oxide with a concentration of 4.5 mg/mL. The advantage of using reducing agents is that they can also be employed to functionalize graphene oxide by bonding with NH_2 groups while reducing it in the solution [50]. By means of Raman spectroscopy, a relationship between the intensities of the D and G bands of approximately 1.23 was determined, which implies a greater ordering of the graphene nanosheets.

In general, for both methods reported, the electrical conductivity values were very low, as well as the thermal conductivity, where the values of 0.05 W/mK are much lower than those of non-porous graphene (3000 W/mk) [47-48, 50].

The cross-linking method consists of the application of sol-gel for the crosslinking of the graphene structure using the controlled polymerization of a precursor such as poly vinyl alchohol, ammonia, etc. [1]. This method was developed in the first instance by Worsley et al., [51], who used the sol-gel for the controlled polymerization of Resorcinol (R) and formaldehyde (F), using sodium carbonate ($NaCO_3$) as a catalyst (C) in an aqueous graphene oxide suspension. A molar ratio of R:F of 1:2 was used, the concentration of graphene oxide was 1% wt., and for the catalyst, a molar ratio of R:C of 200:1 was used. This solution was treated at 85 °C until gelation was produced, and further, the solvent was replaced with acetone. Such a process is required to prevent high stress surface water to create excessive pore shrinkage upon supercritical cooling. Finally, supercritical CO_2 was used to form the aerogel structure. The solid was further taken to pyrolysis at a temperature of 1050 °C in a nitrogen atmosphere, where the graphene oxide resulted in cross-linked graphene with covalent bonds between the different nanosheets. The Resorcinol-Formaldehyde resulted in carbon nanoparticles that acted as gaskets for the different graphene nano-sheets. Due to this reason, these nanostructures result in macroscopic materials with electrical conduction properties several orders of magnitude larger ($1x10^2$ S/m) than in traditional graphene aerogels (2.5-5 $x10^{-1}$ S/m). In the same way, thanks to a much more porous structure due to less stacking, the aerogels manufactured by this technique are much less dense, reaching a value of about 10 mg/cm^3, thus obtaining specific surfaces of 548 m^2/g, which is close to that obtained in more complex methods such as CVD. These properties are improved in subsequent studies by varying the percentages of RF in gelation, having a specific surface of 1199 m^2/g and 762 m^2/g in percentages of 0 and 2% wt., respectively, this along with a notable decrease in the electrical conductivity, possibly due to less crosslinking [52].

Meng et al., [53] proposed an improvement of the method reported by Worsley [51], where an alkaline-treated aqueous solution of graphene oxide was used (sodium hydroxide). This approach allowed not only the functionalization of the graphene oxide particles by removing epoxy and alkoxy groups [54], but also its reduction in a slight manner, as confirmed by Raman spectroscopy. After the treatment of the GO solution, the same steps were carried out as in the previous process. This process resulted in the formation of an aerogel with better properties, and dependent on the initial concentration in the alkaline-treated GO solution, having for a maximum concentration of 10% by weight, a density of 0.189 g/cm^3, a specific surface of 762 m^2/g which is a much higher value than that reported by Worsley [51] and which is comparable to that of a CVD aerogel, the electrical conductivity is also greatly improved, reaching values of 528.2 S/m which is five times higher than the previous case.

The template direct assembly, also called ice-templating assembly is a method in which a colloidal solution of reduced graphene oxide is formed, in such a way that it is homogeneous and totally dispersed in the aqueous solution. Later on, this solution may or may not be gelled by the action of other elements that can serve as auxiliaries in this process, such as PVA [55]. This solution is cooled at a high rate by an external agent that is usually liquid nitrogen, so that later, through a lyophilisation process, the solvent (s) are sublimated, leaving graphene in a more orderly and homogeneous structure than in a direct assembly [17, 56, 57]. During the freezing process, the solvent solidifies at a rate determined by the cooling rate, causing the suspended particles to migrate to the interstices of the crystals [58]. For this reason, controlling the size of the solvent crystals in the freezing step is directly related to certain morphological properties of the prepared aerogel, such as the size of the pore and their homogeneous distribution. At high cooling rates, such as the case of the use of liquid nitrogen as a freezing medium, the pore size is inconsistent, due to the dendritic nucleation of the solvent in contact with the walls where the heat is evolved. [58, 59]. By having a constant and controlled cooling rate, it is possible to obtain a greater order and homogenization of the structure, this rate being directly related to the shape and size of the pore [60].

The freezing temperature is important since it will also determine the size of the crystals, analogously to the growth of grain in solidification, a lower temperature will result in monoliths with smaller grains, which, in the case of porous materials, will influence the value. Young's modulus, in addition to being related to its strength and toughness [61]. Accelerated and controlled cooling can be of great importance since it leaves the ordering that the colloidal solution presented to a large extent frozen, this greatly helps the nanostructure to be homogeneous [55]. Vickery et al., [55] obtained a porous material with a uniform cavity size, where graphene walls of average thickness of 0.35 nm can be seen, use submergence rates in liquid nitrogen of between 2 to 5 .9 mm/min.

One of the main problems when using graphene synthesis methods employing a reducing agent is that the structure does not have very high levels of sp^2 hybridization [55]. The above said together with a clear existence of topological defects and vacancies, make the resulting material of lower quality than that of graphene produced by other methods [46, 55]. For this reason, a method is required to improve the nanostructure of graphene sheets, as proposed by Wang et al., [62], who used the same method. The structure was further impregnated with a KOH solution and thermal reduction was applied. The result was a mesoporous nanostructure with a very high specific surface 990 m^2/g and a low density, reaching values of 8 mg/cm^3. This approach shows that the combination of simple techniques could be applied to reach even higher values of specific surfaces compared to more sophisticated methods.

Carbon Nanotube Aerogels: General Aspects and Synthesis

Carbon nanotubes (CNT), like graphene, are allotropic forms of carbon that could be described as a rolled graphene sheet, thus forming a one-dimensional (1D) nanometric material, whose diameter can vary between 0.7 nm and 15 nm as depending on the synthesis method, this fact being observed for the first time by its discoverer Sumio Iijima in 1991 when he made an arc deposition of carbon soot [63].

This type of materials can be classified through two different criteria, the first is depending on the number of layers that this material has, as a single layer (SWCNT) or multilayer (MWCNT), each with advantages and disadvantages that will be briefly described later. The second classification criterion refers to the way in which the graphene sheet is rolled, which can be chiral, armchair, and zigzag [64].

SWCNTs (Single Walled Carbon Nanotubes) as its name indicates, is a type of nanotube that has only one layer of graphene wound throughout its structure, so the bonds that govern throughout it are covalent, providing of great rigidity, resistance, and conductivity. The synthesis forms for this type of CNT are complicated and obtain up to a maximum of 50% wt. of purity by chemical vapour deposition, and 80% wt. in arc discharge. It is foldable and difficult to functionalize, requiring special methods and conditions. For the synthesis of volumetric materials, it requires an appropriate control in addition to the presence of catalysts that can contaminate the final object [64, 65].

MWCNTs (Multi Walled Carbon Nanotubes) have several concentric layers of nanotubes joined by Van der Waals forces, they exhibit a lower conductivity than SWCNTs. They are easier to synthesize in bulk form, a task that can be done without the presence of a catalyst. The synthesis method applied is primarily CVD. Something that should be mentioned is that these types of structures, have diameters of up to 15 nm, thus, they are larger than SWCNT, and for larger sizes they are called nanocarbon fibers, which have a structure more similar to a bunch of sheets of graphite [64, 66].

Carbon nanotubes are one of the most resistant existing elements, for this reason they generate great interest in a myriad of applications. Some of these properties are listed in Table 3.

These materials are generally synthesized on substrates, classically silicon, and grow from seed points that act as nucleation points from where they begin to grow upwards, thus forming the well-known "CNT forests." Synthesis methods are varied and include arc discharge, laser vaporization, chemical vapour deposition (CVD). These are formed, in the case of CVD, from carbon-rich chemical precursors such as methane, CO, alcohol, etc. [64].

Due to the excellent properties, much interest has been generated to synthesize these materials on a larger volumetric scale, for this reason their production in the form of aerogels is required to take advantage of them in various applications. Aerogels derived from carbon nanotubes are materials with a high specific resistance, being able to resist up to 8000 times their weight [67]. They are ultra-light materials with densities of 0.001 g/cm^3, with a high specific surface area, with hydrophobic properties and with good electrical conductivity in some cases. They are used in various applications such as environmental remediation, electronics applications, biotechnology, batteries, etc.

Table 3. Main physical properties of CNTs

Property	Value	Reference
Density [g/cm^3]	1.3	[64]
Young Modulus [Tpa]	1	[64]
Tensile Strength [Gpa]	63	[64]
Stiffness [Tpa]	1	[64]
Electrical C. [MS/m]	0.01-1	[64]
Thermal C. [W/mK]	3000	[64]

As in the case of graphene, it is possible to start from precursors that form the nanotubes in a nanostructure, or by using these in a dispersion that will gel and subsequently be treated to form the desired architecture. All the techniques presented below have their peculiarities, advantages and disadvantages that make them ideal for certain specific applications.

CVD synthesis is a method in which a carbon-rich substance is used to, through an evaporation process, it is deposited on a surface. For this purpose, a substance rich in iron is usually used as a nucleating element, which is usually ferrocene. Carbon nanotubes tend to grow uniformly, and seemingly indefinitely long [68]. To introduce a random order, dichlorobenzene is used, which is a substance that has proven to disturb the aligned growth of carbon nanotubes [69]. To obtain the desired shape, the CNT aerogel is drawn mechanically, directing the gas flow to the desired location. Synthesis parameters determine certain properties of the final material. For example, Li et al., [68] used ethanol as a precursor in conjunction with ferrocene (0.23-2.3% wt.) and thiophene (1–4% wt.) as catalysts at an injection rate of 0.08-0.25 ml/min in a temperature range from 1050 °C to 1200 °C, managing to obtain CNT nanostructures with a purity of up to 90%. However, the resistance and conductivity of the products were scarce. Better results can be obtained by varying certain parameters, such as deposition surface and reaction temperatures, with 860°C and a quartz surface, as shown by Gui et al., [70] who synthesized a sponge of MWCNT with better electrical conductivity (1.7 S/cm). Its thermal conductivity was 0.15 W m^{-1} K^{-1} and it had a porosity greater than 99% which allowed it to absorb different oils with a ratio of 80 to 180 times its weight. Likewise, its specific surface exhibits value between 300 and 400 m^2/g. This type of aerogels can be improved thanks to the inclusion of doping elements or by combining them with other types of substances. Deposition rates are usually 2-3 mm/h, regardless of the type of substrate used for the formation of these [69].

The nanostructures formed by this method usually have drawbacks directly associated with the method, that is, CVD requires a lot of energy in addition to specialized facilities, which makes the final material more expensive, apart from problems derived from the use of toxic vapours that can harm the health of the people who operate this equipment [1].

To improve the bonds between nanotubes, and consequently the mechanical and electrical properties of the aerogel, these nanostructures are often doped with boron or nitrogen [69]. For this purpose, precursors such as toluene and triethylborane are used in conjunction with ferrocene in the form of aerosols.

Another way to synthesize these nanostructures is by using CVD, but combined with pre-made templates [69], which can be porous nickel or ZnO, as in the case of graphene ones [44]. This way, it is possible to have much more homogeneous structures, and although with high porosity, they have a higher ordering than those that depend only on the random generation of carbon nanotubes.

Sol-gel method aims to form a CNT aerogel, through sol-gel. However, this task is complicated due to the insoluble nature of CNTs. For this reason, surfactants are used in such a way as to promote a better inclusion of nanotubes in the sun. After this, gelling is induced to maintain the nanotubes in a dispersed form. With the synthesized gel, it is possible to remove the surfactant water through continuous bath, then, to remove the water, as in the classic synthesis of aerogels, where a solvent change is carried out, generally using ethanol for this step. Finally, the solvent is removed through lyophilisation or supercritical evaporation, leaving a three-dimensional structure. The aerogels produced by this technique have good conductive properties, in addition to a very high specific surface (900-1300 m^2/g) [71], these values varying according to the concentration of CNT used in the sun. As a surfactant, it is common to use sodium dodecyl benzene sulphonate (SDBS) through sonification. While the properties allow these aerogels to compete as candidates for many applications, they present the problem of severe brittleness, so they must be handled with care. To solve this, it is usually combined with polymeric matrices, which, although they improve flexibility, sacrifice their electrical conductivity and, to a lesser extent, the specific surface.

Ice-templating method also uses an aqueous and dispersed solution of CNT, but unlike the sol-gel transition, it is frozen at a certain temperature and rate [1]. These, in addition to factors such as the size of the colloids, pH of the solution and concentration of the precursors, will determine the way in which the ice crystals grow in the solution, and consequently, govern the size, shape and distribution of pores [72, 73]. The aerogels obtained by this method have high specific surfaces (400 m^2/g), very low densities (4-15 mg/cm^3) and good electrical conductivities (0.032 - 0.67 S/cm) [74]. Since freezing occurs from the outside to inside, it is common to observe a gradient of pore sizes along the thickness of the formed aerogel. This, together with the tendency of crystals to disperse dispersed colloidal particles, can affect the final properties of the carbon nanotube aerogel [1]. For this reason, the properties, mainly mechanical, are usually improved using particles that will be in the colloid and that will serve as an emulsifier, such as PVA, and constitute the areas where the nanotubes will tend to cluster. This helps to have better crosslinking in the aerogel network, to improve its electrical properties [72]. For example, Xia et al. [72] synthesized CNT aerogels by ice-templating with the use of emulsifiers and obtained better gas permeability and greater resistance, in addition to reaching smaller pore sizes. However, the specific surface area decreases from 358 m^2/g by ice-templating, to 290 m^2/g when emulsifiers were employed. For this reason, the density also increased by 50%. The electrical and thermal conductivities were improved, possibly due to the greater crosslinking in this type of synthesis.

Nanodiamond Aerogels: General Aspects and Synthesis

Nanodiamonds are, like graphene, fullerenes, CNT and others, allotropic forms of carbon that are in a metastable state. Discovered since the 1960s in the USSR, but known in the West since the 1980s, they were first synthesized through detonation tests [75]. In these tests, the explosive material, which is rich in carbon, reaches very high temperatures and pressures over short periods of time, thus the carbon resulting from the explosion is strongly compressed and forms this type of nanoparticles [75]. Today, the synthesis of nanodiamonds through the detonation method is still used and it is carried out in controlled chambers in inert atmospheres. The average size of particles obtained by this method ranges from 4 nm to 5 nm [76].

In addition to such "traditional" method, it is possible to synthesize nanodiamonds by other methods including laser ablation, high-energy ball milling, high pressure - high temperature (HPHT), plasma assisted chemical vapor deposition, autoclave synthesis from supercritical fluids, chlorination of carbides, ion irradiation of graphite, electron irradiation of carbon onions, and ultrasound cavitation. The yield of nanodiamonds from the precursor explosion is about 75% wt. nanodiamonds, the rest being pollutants that are mostly carbon traces that were not able to be converted to this type of material (4-10% wt.). In addition, there is a presence of up to 5% wt. of metals [77].

The main drawback of these nanoparticles is not their purity but rather their tendency to form very large agglomerates. To purify the product, this is subjected to different chemical processes and thus, the metals and carbon residues are removed by employing mainly nitric acid, sulfuric acid, or a mixture of both [77]. After this process, it is necessary to carry out a final treatment with ozone, in such a way that these nanoparticles are decorated with some functional groups that allow them to form a colloidal suspension [76] and to be employed for subsequent synthesis in the form of aerogel.

On the other hand, nanodiamond aerogels are materials that have porous structures with cavities in the order of nanometers. They have inherent properties, such as low densities (200 mg/cm^3), and low toxicity and biocompatibility, thanks to which they are studied as possible means for drug delivery. They also exhibit high specific surfaces, which makes them attractive for application in photocatalysis, photonics and astrophysics [78-79]. One of the main potential qualities of this type of aerogels relies on its optical properties, which would allow it to be an alternative for the manufacture of anti-reflective coatings.

In addition to the classical properties of a carbon aerogel, this type of materials, being derived from diamond, also presents nitrogen vacancy centers, which are point defects in the crystal lattice consisting of a pair closest neighbor of a nitrogen atom and replacing a carbon atom, and a lattice vacancy. These can be further exploited as units in quantum computing. The advantage of these defects is that they can be manipulated by magnetic and electric fields at room temperature.

The synthesis of this type of materials is a real challenge for the community focused on materials. The traditional methods usually start from the synthesis of nanodiamonds embedded in a three-dimensional structure, which allows to have the desired porous element.

In the high pressure – high temperature synthesis method, the carbon nanostructured precursor is generally an amorphous carbon aerogel, so that, by using a diamond anvil cell and by applying heat at a very high temperature, such precursor transitions to the diamond state, thus, keeping the nanostructure porous. Pauzauskie et al., [78] managed to obtain a porous material composed mainly of nanodiamonds by this method. For this purpose, a carbon aerogel with a density of 0.04 g/cm^3 was used, placed in a cell where it was filled with supercritical neon at very high pressure (22000 psi). This is done because the diamond anvil cell performs a compression in the order of 21–25 GPa, so that the nanostructure is not completely destroyed, and the internal pressure of the aerogel must be equal to the external pressure exerted through the cell. Such conditions require a very delicate and precise control to keep the pores intact as far as possible. Neon is selected due to its transparency, which is very useful when direct heating to the aerogel. For this purpose, a powerful laser is used, which heats the aerogel to a temperature of 1580 K. Another advantage of using this cell is that due to its transparency so that it is possible to perform spectroscopy measurements (X-ray and Raman) to study the transition of this material to the diamond crystal structure. After the experiment is carried out, a new peak is observed in the Raman spectrum, as well as the formation of rings in the diffractogram carried out in conjunction with TEM microscopy, indicating the transition. This is confirmed again when comparing the X-ray absorption spectrum with that of nanodiamonds obtained by detonation, where similar figures were observed [78]. Using the described method, it was possible to obtain a nanodiamond aerogel with sizes from 2.5 to 100 nm.

As with other carbon aerogels, it is also possible to obtain a nanodiamond aerogels by the sol-gel method. In this approach, the nanodiamonds are suspended in colloidal form in a Resorcinol-Formaldehyde (R-F) solution, and by means of gelling and subsequent lyophilization or supercritical evaporation, the required aerogel is obtained. With the purpose of obtaining nanodiamonds that can be dispersed in a solution prior to the R-F gelling process, these must be endowed with functional groups that contain oxygen (hydroxyls, carboxyl, etc.), for this an oxidation process is carried out as mentioned above. For this, tube furnaces are usually used at a suitable temperature to remove all the remaining amorphous carbon in the nanodiamonds. Manandhar et al., [79] synthesized nano-diamond aerogels by the oxidation of ND90, dynalene, in a tube furnace at 450 °C for 8 hours. Subsequently, 17.5 mg of these were used to form a dispersion with acetonitrile as a solvent through sonification. Later, resorcinol and formaldehyde were added for gelling purposes at room temperature. Finally, the gel was subjected to solvent exchange with ethanol prior to supercritical drying with CO_2. A crystalline nanostructure of cross-linked nanodiamonds through covalent bonds between them is obtained, and by examination with FTIR and X-ray diffraction, in addition to surface area tests, the presence of hydroxyl groups was confirmed and in respect to the specific surface, high values were obtained, reaching 589 m^2/g, a density of about 151 mg/cm^3. Likewise, by using electron microscopy, it was possible to determine that the size of nanodiamonds was in the order of 5 nm.

Nanocarbon Aerogels Applied in Supercapacitors

To promote the use of environmentally friendly technologies, it is necessary to develop elements that allow their viability, and one of the main challenges in this field is that of storing electrical energy, not only for later use but also for applications such as converters in large power plants [80]. Its use has not only been limited to this field, but it was also applied to that of electric vehicles, battery improvement [81], and above all to its use in renewable energies whose sources have fluctuations in the amounts of energy they deliver.

Since their invention in 1957 by General Electric, supercapacitors have had several applications where high storage of electrical energy is required, commonly a common capacitor has capacitance values in the order of millifarad, while supercapacitors show values of various orders of magnitude higher, even reaching hundreds of Farads per gram of material thanks to the high specific surfaces. In addition to this, they have properties that give them a long useful life and very high energy densities, which in the case of graphene aerogels show values of 300 to 5000 kW/kg, which is 5 to 10 times higher when compared to lithium batteries [82].

Its use is limited mainly by its high cost due to the synthesis techniques used, which for example in the case of graphene aerogel, is usually made by chemical vapour deposition, this due to the large specific surfaces reached, in addition to the fact that its electrical conductivity improves substantially. For example, for the case of graphene aerogels, Luan et al., [83] obtains supercapacitors with values of 160 F/g by hydrothermal reduction, Zhang et al., [84] reported capacitances of 231 F/g by the cross-linking method, and Stoller et al., [85] obtained aerogels with a capacitance of 135 F/g. Hao et al., [81] proposed a nitrogen doping to a graphene aerogel previously obtained by the crosslinking of chitosan and carried out a subsequent activation treatment with KOH. This aerogel exhibited a specific capacitance of 197 F/g, and from the charge and discharge tests, it resulted to have a retention value greater than 90% even after 10000 cycles. Likewise, by R-F polycondensation with KOH activation, Li managed to obtain surface supercapacitors and specific capacitances of 656 m^2/g and 183.6 F/g, respectively [86].

Nanocarbon Aerogels Applied in Hydrogen Storage

As we become more aware of the impact on the planet that the use of fossil fuels has, solutions appear that could greatly alleviate our dependence on these fuels. Hydrogen is a fuel that for several years promises to be the definitive solution to the problem of fuels, it has a good energy capacity and combustion waste that is totally harmless to the environment, since the oxidation of this fuel generates water [87].

One of the main problems with the use of this fuel is the lack of efficient forms of transport and storage. A solution to this drawback was presented as storage options in liquid or metallic hydride form (AlH_4, NH_2, BH_4) [88], solutions that lead to other problems such as lack of compatibility with current systems in the case of vehicles, or very high storage costs.

The mechanism by which hydrogen is stored in the carbon aerogel nanostructures is physisorption, which is an interaction due to Van der Waals forces between the storage fluid and the container [88], this force is very weak, which depends directly on the contact surface, which in this case is represented by the specific surface. It is also affected by temperature, where for lower temperatures, the physisorption is higher and therefore there will be a greater amount of retained fluid [80].

To achieve acceptably efficient amounts of hydrogen storage, high surface areas are required. In this respect, Biener et al., [80] found a ratio of 500 m^2/g for each 1% wt. of hydrogen retained, which is why surfaces greater than 2000 m^2/g are required, otherwise very low storage temperatures will be required (77 K) [89], which makes this difficult, this is because the porous carbon structures have low surface binding energies, approximately 6 kJ mol^{-1} of H_2 [80]. To obtain high specific surfaces, the carbon nanostructures are activated by using a chemical agent (the most used are KOH, H_3PO_4, $ZnCl_2$, which react with the aerogel at a certain temperature for each substance in an inert atmosphere). For example, Wang and Kaskel [90] obtained carbon structures with a specific surface area of 3000 m^2/g by using KOH at reaction temperatures of 650 °C.

Nanocarbon Aerogels Applied in Water Desalination

There are currently alarming waves of intense drought in various parts of the world, the most recent in Mexico, which has suffered from an intense drought since July 2020, the same that has affected 87.5% of the territory [91]. For these reasons, it is necessary to obtain water for consumption from alternative sources, one of the most used being that of desalinated water. For this purpose, one of the most promising techniques is capacitive deionization, which is an electrochemical technique that removes excess ions dissolved in an aqueous solution using electrodes with a double electrical layer [92], through the application of a relatively low potential difference (0.6-2.0 V DC) [93].

Carbon-based aerogels, being materials with a high specific surface area, represent an alternative as a material for the construction of double-layer electrodes, in which one of the layers is positively charged, attracting negative ions, also for the case. reverse. Jung et al., [94] achieves the synthesis of an aerogel based on R-F, which shows a high good electrical conductivity and specific capacitance (220 F/g), obtaining a desalination with 97.6% at 1.7 V. This technique is not only useful for the elimination of common salt, but also of calcium salts since it takes advantage of the fact that these ions are separated in the solution.

Another technique used for the desalination of water is the generation of steam through sunlight, in this technique use is made of the light absorption capacities of carbon-based materials. Storer et al., [95] used an aerogel based on graphene oxide nano-sheets reduced by the hydrothermal method, which were assembled with rice straw to give it more stability, after which pyrolysis was carried out, and it was determined that the nanostructure absorbed 96-97% of the light solar, thus being able to evaporate water at a rate of 2.25 kg m^{-2} h^{-1}. The wastewater obtained was characterized and it was determined that its concentration wasof 0.37 ppm.

Nanocarbon Aerogels Applied in Water Pollutant Removal

Within environmental remediation, one of the greatest challenges is to repair the ecosystems affected by problems such as contamination of aquifer sources resulting from the presence of pollutants that can range from contamination by chemical elements used as colorants in the textile industry, the presence of heavy metals in groundwater, or the presence of oils resulting from the spill of petroleum derivatives, accidents that unfortunately have occurred many times and whose impact is incalculable. In addition to various efforts to repair this either chemically, such as water treatment, or mechanically with the use of barriers and other tools, the separation and absorption of these has proven to be a very important remediation technique in recent years.

The absorbents employed to this end should ideally be chemically inert elements, environmentally friendly, with a high specific surface, and with hydrophobic properties, in such a way that they serve at the same time as a separating element between the water and the pollutants present, for these reasons the carbon-based materials have proven to be a good alternative for the manufacture of absorbent elements for this purpose.

Aerogels and sponges made from carbon nanotubes are a very attractive option, aerogels based on graphene, as well as their use with metallic nanoparticles, have shown excellent properties for absorbing oil [96]. For example, Wu et al., [97] showed the synthesis of a graphene aerogel by reduction with copper, which exhibited good absorptive properties, removing 20 to 40 times its weight with different oils. In the same way, it showed a great behaviour in the dye degradation such as methylene orange, used for modelling dye removal experiments, which was due to the catalytic and photocatalytic properties of the copper compound and graphene aerogel. Another parameter for which these materials stand out is their absorption rates, where with a simple reduction method it is possible to obtain graphene aerogels with high rates [98].

Even better results are obtained with materials with higher hydrophobicity, as shown by Gui et al., [99] who synthesized CNT sponges using CVD method which shown super hydrophobicity with contact angles of 156 °, managing to absorb 80 to 180 times their weight through tests with different organic solvents.

Hybrid aerogels of graphene and nanotubes can also be a good option due to nanostructures with greater number of pores, this is due to the combination of the great hydrophobicity of CNT and flexibility of graphene, which allows its reuse without greatly losing its absorptive properties. Kabiri et al., [49] synthesized a hybrid aerogel of graphene and CNT by reduction with iron, which, by combining it with vacuum absorption, showed ability to continuously remove 28 L of oil per gram of aerogel in 1 hour in non-turbulent media. These results indicate a tremendous potential of this type of material for the removal of pollutants and purification of the water.

The heavy metals that are dissolved in the form of ions in aquifer wells and groundwater, which are the majority sources in many towns and cities, can cause a great prejudice to health. Among these, lead, cadmium, mercury can be mentioned. The possible bond between these ions and a porous absorbent material is governed mainly by electrostatic forces [100], something that is very beneficial for some allotropic forms that may contain oxygen groups on their surface, this can be beneficial if using aerogels with high graphene oxide content, or hybrids of graphene and CNT. For example, Zhuyin et al. [101] obtained a hybrid aerogel by reducing graphene oxide with CNT and ethylenediamine, that presented good binding capacities with metal ions, having 104.9 mg/g (0.51 mmol/g) for Pb^{2+}, 93.3 mg/g (0.46 mmol/g) for Hg^{2+}, 64 mg/g (0.59 mmol/g) for Ag^{1+} and 33.8mg/g (0.53 mmol/g) for Cu^{2+}.

Nanocarbon Aerogels Applied in Catalysis

Catalysis is an area of potential application for carbon aerogels since they are porous elements with a large surface area, this added to the possible improvement with doping elements that improve their electrical conductivity makes them good candidates to be considered to replace electrodes. of traditional materials such as platinum in catalytic reactions.

Advances have been made in the field of oxygen reduction reactions (ORR) through the use of iron-doped graphene aerogels, as shown by Yin et al., [102] who synthesized an aerogel with electrolytic properties capable of competing with Pt / C electrodes by reducing graphene oxide in a solution with iron nitride, and achieved higher current densities for the same potential ranges of platinum electrodes, and oxygen depletion peaks at voltages similar to platinum (-0.12 mV).

On the other hand, Zhang et al., [103] applied the chemical reaction reactions of p-nitroaniline, and prepared a graphene aerogel with platinum nanoparticles, achieving almost complete reduction of p-nitroaniline in the presence of $NaBH_4$ in 10 minutes, which was further confirmed by UV-Vis spectroscopy. Similar results were also obtained but for the reduction of 4-nitrophenol, also in the presence of $NaBH_4$ by using graphene aerogels with carbon spheres [104], and copper nanoparticles [97].

Likewise, the hybridization of carbon nanotube aerogels with metals is possible and thus, enhancing the reaction speed in energy cells such as lithium-ion batteries by the use of CNT and palladium sponges [105]. Similar to graphene aerogel electrodes, the CNT ones can be doped with nitrides to be used as an alternative to palladium electrodes in ORR reactions, obtaining higher current densities at operating voltages less than -0.4 V (vs Ag / AgCl) [69].

Nanocarbon Aerogels Applied in Electronics

One of the advantages of using materials with a high surface area and with good conductive properties is that the detection window is much wider, which increases the general sensitivity of the system. Such aspect is especially useful in the use of this type of oriented materials to develop sensors and actuators. The use of carbon-based aerogels can also apply as the structure where the material that will perform the function is established, thus obtaining a synergy with the union of the best properties of both components. For example, CVD-synthesized graphene aerogels can detect dopamine with a high sensitivity, allowing it to detect concentrations as low as 25 nM [45].

The principle of detection of this type of sensors consists in the change of resistivity that they present thanks to the absorption of a particular substance. For this reason, it is necessary that these materials exhibit a good conductivity so that their sensitivity is high enough to be taken in account. This is the case of carbon nanotube aerogels, which, thanks to their high conductivity and absorption capacity for certain functional groups, can detect the presence of organic vapours such as toluene, methanol, ethanol, etc. However, the addition of another component is often necessary to increase the selectivity among all possible substances present in the air that can distort the measurements [106].

Long et al. [107] obtained an ultra-sensitive aerogel by using graphene aerogel and MoS_2, that could detect NO_2 with a detection limit of 50 ppb, as it used the high surface area of the aerogel as a scaffold. Moreover, it also showed improved selectivity of NO_2 against H_2 and CO due to MoS_2.

It is possible to obtain architecture materials with ability to detect different types of substances, e.g., hydroquinone and o-dihydroxybenzene by the combination of this type of aerogels with other types of materials, e.g., gold particles that are used in a nitrogen-doped graphene aerogel [108].

Thanks to the great flexibility and versatility of graphene aerogels, these can be combined during their synthesis with magnetite nanoparticles or another material that responds to electromagnetic fields, for this reason they can exhibit up to 52% deformation induced by magnetic fields, which makes it a candidate to be used as an actuator or micro-switch [109].

In addition to the use of aerogels as sensors and actuators, it is also possible to use intrinsic properties of carbon in shielding against magnetic interference. For example, graphene aerogels in combination with cellulose nanofibers have demonstrated ultra-high specific shielding capabilities in addition to its high flexibility allowing this capacity to be regulated through mechanical compression [110].

Derived from the properties of flexibility and electrical conductivity, added to a decent specific mechanical resistance, it is possible to develop elements that respond to an electrical potential inducing an elastic contraction, as is the case of artificial muscles by employing a combination with other materials. For example, Foroughi et al., [111] developed a spandex core compound surrounded by CNT aerogel, which was spun until obtaining a fabric that responded to electrothermal heating and achieved longitudinal contractions of up to 33%, with an associated work of 0.64 kJ/kg, which emulates a muscle.

Conclusion

Carbon nanomaterials including carbon nanotubes, graphene and nanodiamonds have gained increasing research interest in the last years. Large volume of research has been dedicated to the synthesis of such materials as three-dimensional architectures and thus to exploit their properties as nanoscale also at larger ones.

This chapter focused on presenting the synthesis methods applied to fabricate aerogel forms from carbon nanotubes, graphene and nanodiamond material sources. Varying characterisctics of the carbon nanomaterial sources have been highlighted. The recent results in varying fields of application of such aerogels have been briefly presented.

The results obtained on this topic demonstrated the viability of such nanomaterials to obtain performant structures in many fields including water treatment, electronics, catalysis or energy storage. Nevertheless, a combination of the synthesis method and the proper balance of the characteristics of nanocarbon source material such as surface area, functional groups, degree of reduction etc. is necessary to tailor the performance of their three-dimensional structures.

Acknowledgments

N. M. R. L. is grateful for the financial support from Escuela Politécnica Nacional.

References

[1] Araby, S., A. Qiu, R. Wang, Z. Zhao, C. H. Wang, and J. Ma, "Aerogels based on carbon nanomaterials," *J. Mater. Sci.*, vol. 51, no. 20, pp. 9157–9189, 2016, doi: 10.1007/s10853-016-0141-z.

[2] Zhao, N., et al., "Hierarchical porous carbon with graphitic structure synthesized by a water soluble template method," *Mater. Lett.*, vol. 87, pp. 77–79, 2012, doi: 10.1016/j.matlet.2012.07.085.

[3] Lin, Y., G. J. Ehlert, C. Bukowsky, and H. A. Sodano, "Superhydrophobic functionalized graphene aerogels," *ACS Appl. Mater. Interfaces*, vol. 3, no. 7, pp. 2200–2203, 2011, doi: 10.1021/am200527j.

[4] Fecht, H., K. Brühne, and P. Gluche, *Carbon-based Nanomaterials and Hybrids*. 2016.

[5] Serp, P., *Carbon*, vol. 7. 2013.

[6] Gupta, S., and N. H. Tai, "Carbon materials as oil sorbents: A review on the synthesis and performance," *J. Mater. Chem. A*, vol. 4, no. 5, pp. 1550–1565, 2016, doi: 10.1039/c5ta08321d.

[7] Zhan, W., L. Gao, X. Fu, S. H. Siyal, G. Sui, and X. Yang, "Green synthesis of amino-functionalized carbon nanotube-graphene hybrid aerogels for high performance heavy metal ions removal," *Appl. Surf. Sci.*, vol. 467–468, pp. 1122–1133, 2019, doi: 10.1016/j.apsusc.2018.10.248.

[8] Nardecchia, S., D. Carriazo, M. L. Ferrer, M. C. Gutiérrez, and F. Del Monte, "Three dimensional macroporous architectures and aerogels built of carbon nanotubes and/or graphene: Synthesis and applications," *Chem. Soc. Rev.*, vol. 42, no. 2, pp. 794–830, 2013, doi: 10.1039/c2cs35353a.

[9] Olszowska, K., et al., "Three-dimensional nanostructured graphene: Synthesis and energy, environmental and biomedical applications," *Synth. Met.*, vol. 234, no. August, pp. 53–85, 2017, doi: 10.1016/j.synthmet.2017.10.014.

[10] Novoselov, K. S. et al., "Two-dimensional atomic crystals," *Proc. Natl. Acad. Sci. U. S. A.*, vol. 102, no. 30, pp. 10451–10453, 2005, doi: 10.1073/pnas.05028 48102.

[11] Novoselov, K. S., et al., "Electric Field Effect in Atomically Thin Carbon Films," *Science (80-.).*, vol. 306, no. 2, pp. 666–669, 2004, doi: 10.1016/j.medcli. 2015.04.005.

[12] Goodenough, J. B., and K. S. Park, "The Li-ion rechargeable battery: A perspective," *J. Am. Chem. Soc.*, vol. 135, no. 4, pp. 1167–1176, 2013, doi: 10.1021/ja3091438.

[13] Tan, Z., G. Chen, and Y. Zhu, "Carbon-Based Supercapacitors Produced by Activation of Graphene-surporting materials," *Science (80-.).*, vol. 1,

no. June, pp. 211–225, 2015, [Online]. Available: http://science.sciencemag.org/.

[14] Simon, P., and Y. Gogotsi, "CarbonÀElectrolyte Systems," vol. XXX, no. Xx, 2011.

[15] Tiwari, J. N., et al., "Reduced graphene oxide-based hydrogels for the efficient capture of dye pollutants from aqueous solutions," *Carbon N. Y.*, vol. 56, pp. 173–182, 2013, doi: 10.1016/j.carbon.2013.01.001.

[16] Lee, B., et al., "Carbon nanotube-bonded graphene hybrid aerogels and their application to water purification," *Nanoscale*, vol. 7, no. 15, pp. 6782–6789, 2015, doi: 10.1039/c5nr01018g.

[17] Nasir, S., M. Z. Hussein, Z. Zainal, and N. A. Yusof, "Carbon-based nanomaterials/allotropes: A glimpse of their synthesis, properties and some applications," *Materials (Basel).*, vol. 11, no. 2, pp. 1–24, 2018, doi: 10.3390/ ma11020295.

[18] Wan, W., et al., "Graphene-carbon nanotube aerogel as an ultra-light, compressible and recyclable highly efficient absorbent for oil and dyes," *Environ. Sci. Nano*, vol. 3, no. 1, pp. 107–113, 2016, doi: 10.1039/c5en00125k.

[19] Alshrah, M., M. P. Tran, P. Gong, H. E. Naguib, and C. B. Park, "Development of high-porosity resorcinol formaldehyde aerogels with enhanced mechanical properties through improved particle necking under CO2 supercritical conditions," *J. Colloid Interface Sci.*, vol. 485, pp. 65–74, 2017, doi: 10.1016 /j.jcis.2016.09.030.

[20] Wu, Z. Y., et al., "Dyeing bacterial cellulose pellicles for energetic heteroatom doped carbon nanofiber aerogels," *Nano Res.*, vol. 7, no. 12, pp. 1861–1872, 2014, doi: 10.1007/s12274-014-0546-4.

[21] Zhang, J., G. Chen, Q. Zhang, F. Kang, and B. You, "Self-Assembly Synthesis of N-Doped Carbon Aerogels for Supercapacitor and Electrocatalytic Oxygen Reduction," *ACS Appl. Mater. Interfaces*, vol. 7, no. 23, pp. 12760–12766, 2015, doi: 10.1021/acsami.5b01660.

[22] De Nicola, F., et al., "Wetting Properties of Graphene Aerogels," *Sci. Rep.*, vol. 10, no. 1, pp. 1–7, 2020, doi: 10.1038/s41598-020-58860-4.

[23] Wu, S., et al., "Strain Sensors with Adjustable Sensitivity by Tailoring the Microstructure of Graphene Aerogel/PDMS Nanocomposites," *ACS Appl. Mater. Interfaces*, vol. 8, no. 37, pp. 24853–24861, 2016, doi: 10.1021/acsami.6b06012.

[24] Wang, H., et al., "Three dimensional graphene based materials: Synthesis and applications from energy storage and conversion to electrochemical sensor and environmental remediation," *Adv. Colloid Interface Sci.*, vol. 221, pp. 41–59, 2015, doi: 10.1016/j.cis.2015.04.005.

[25] Pei, S., Q. Wei, K. Huang, H. M. Cheng, and W. Ren, "Green synthesis of graphene oxide by seconds timescale water electrolytic oxidation," *Nat. Commun.*, vol. 9, no. 1, pp. 1–9, 2018, doi: 10.1038/s41467-017-02479-z.

[26] Yu, Z., C. Hu, A. B. Dichiara, W. Jiang, and J. Gu, "Cellulose nanofibril/carbon nanomaterial hybrid aerogels for adsorption removal of cationic and anionic organic dyes," *Nanomaterials*, vol. 10, no. 1, pp. 1–20, 2020, doi: 10.3390/nano 10010169.

[27] Shao, X., W. Lu, R. Zhang, and F. Pan, "Enhanced photocatalytic activity of TiO 2 -C hybrid aerogels for methylene blue degradation," *Sci. Rep.*, vol. 3, pp. 1–9, 2013, doi: 10.1038/srep03018.

[28] Worsley, M. A., et al., "Synthesis and characterization of highly crystalline graphene aerogels," *ACS Nano*, vol. 8, no. 10, pp. 11013–11022, 2014, doi: 10.1021/nn505335u.

[29] Kundu, M. K., T. Bhowmik, and S. Barman, "Gold aerogel supported on graphitic carbon nitride: An efficient electrocatalyst for oxygen reduction reaction and hydrogen evolution reaction," *J. Mater. Chem. A*, vol. 3, no. 46, pp. 23120–23135, 2015, doi: 10.1039/c5ta06740e.

[30] Zhu, Y., et al., "Graphene and graphene oxide: Synthesis, properties, and applications," *Adv. Mater.*, vol. 22, no. 35, pp. 3906–3924, 2010, doi: 10.1002/adma.201001068.

[31] Han, M., et al., "Fabricating 3D Macroscopic Graphene-Based Architectures with Outstanding Flexibility by the Novel Liquid Drop/Colloid Flocculation Approach for Energy Storage Applications," *ACS Appl. Mater. Interfaces*, vol. 10, no. 26, pp. 21991–22001, 2018, doi: 10.1021/acsami.8b02942.

[32] Xing, C., et al., "Graphene oxide/black phosphorus nanoflake aerogels with robust thermo-stability and significantly enhanced photothermal properties in air," *Nanoscale*, vol. 9, no. 24, pp. 8096–8101, 2017, doi: 10.1039/c7nr00663b.

[33] Jung, S. M., H. Y. Jung, M. S. Dresselhaus, Y. J. Jung, and J. Kong, "A facile route for 3D aerogels from nanostructured 1D and 2D materials," *Sci. Rep.*, vol. 2, no. i, pp. 1–6, 2012, doi: 10.1038/srep00849.

[34] Hou, Y., et al., "A 3D hybrid of layered MoS2/nitrogen-doped graphene nanosheet aerogels: An effective catalyst for hydrogen evolution in microbial electrolysis cells," *J. Mater. Chem. A*, vol. 2, no. 34, pp. 13795–13800, 2014, doi: 10.1039/c4ta02254h.

[35] Lee, C., X. Wei, J. W. Kysar, and J. Hone, "Measurement of the elastic properties and intrinsic strength of monolayer graphene," *Science (80-.).*, vol. 321, no. 5887, pp. 385–388, 2008, doi: 10.1126/science.1157996.

[36] Sivudu, K. S., and Y. Mahajan, "Mass Production of High Quality Grpahene:An Analysis of Worldwide Patents," vol. 3, no. 2, 2012.

[37] Araby, S., et al., "Electrically and thermally conductive elastomer/graphene nanocomposites by solution mixing," *Polymer (Guildf).*, vol. 55, no. 1, pp. 201–210, 2014, doi: 10.1016/j.polymer.2013.11.032.

[38] Backes, C., et al., "Production and processing of graphene and related materials," *2D Mater.*, vol. 7, no. 2, 2020, doi: 10.1088/2053-1583/ab1e0a.

[39] Yang, J., et al., "Hybridizing graphene aerogel into three-dimensional graphene foam for high-performance composite phase change materials," *Energy Storage Mater.*, vol. 13, pp. 88–95, 2018, doi: 10.1016/j.ensm.2017.12.028.

[40] Zhou, L., Z. Yang, J. Yang, Y. Wu, and D. Wei, "Facile syntheses of 3-dimension graphene aerogel and nanowalls with high specific surface areas," *Chem. Phys. Lett.*, vol. 677, pp. 7–12, 2017, doi: 10.1016/j.cplett.2017.03.076.

[41] Sultanov, F. R., S. S. S. Pei, M. Auyelkhankyzy, G. Smagulova, B. T. Lesbayev, and Z. A. Mansurov, "Aerogels based on graphene oxide with addition of carbon nanotubes: Synthesis and properties," *Eurasian Chem. J.*, vol. 16, no. 4, pp. 265–269, 2014, doi: 10.18321/ectj9.

[42] Kim, K. H., Y. Oh, and M. F. Islam, "Graphene coating makes carbon nanotube aerogels superelastic and resistant to fatigue," *Nat. Nanotechnol.*, vol. 7, no. 9, pp. 562–566, 2012, doi: 10.1038/nnano.2012.118.

[43] Jia, J., et al., "and Fracture Resistance of 3D Interconnected Graphene Foam / Epoxy," *ACS Nano*, no. Xx, pp. 5774–5783, 2014.

[44] Chen, Z., W. Ren, L. Gao, B. Liu, S. Pei, and H. M. Cheng, "Three-dimensional flexible and conductive interconnected graphene networks grown by chemical vapour deposition," *Nat. Mater.*, vol. 10, no. 6, pp. 424–428, 2011, doi: 10.1038/nmat3001.

[45] Dong, X., et al., "3D graphene foam as a monolithic and macroporous carbon electrode for electrochemical sensing," *ACS Appl. Mater. Interfaces*, vol. 4, no. 6, pp. 3129–3133, 2012, doi: 10.1021/am300459m.

[46] Schniepp, H. C., et al., "Functionalized single graphene sheets derived from splitting graphite oxide," *J. Phys. Chem. B*, vol. 110, no. 17, pp. 8535–8539, 2006, doi: 10.1021/jp060936f.

[47] Zhang, L., G. Chen, M. N. Hedhili, H. Zhang, and P. Wang, "Three-dimensional assemblies of graphene prepared by a novel chemical reduction-induced self-assembly method," *Nanoscale*, vol. 4, no. 22, pp. 7038–7045, 2012, doi: 10.1039/c2nr32157b.

[48] Wenchao, W., F. Zhang, S. Yu, R. Zhang, and Z. Ying, "Hydrothermal formation of graphene aerogel for oil sorption the role of reducing agent, reaction time andtemperature," *New J. Chem.*, vol. 2019-Novem, 2019, doi: 10.1039/x0xx00 000x.

[49] Kabiri, S., D. N. H. Tran, T. Altalhi, and D. Losic, "Outstanding adsorption performance of graphene-carbon nanotube aerogels for continuous oil removal," *Carbon N. Y.*, vol. 80, no. 1, pp. 523–533, 2014, doi: 10.1016/j.carbon.2014. 08.092.

[50] Tang, G., Z. G. Jiang, X. Li, H. Bin Zhang, A. Dasari, and Z. Z. Yu, "Three dimensional graphene aerogels and their electrically conductive composites," *Carbon N. Y.*, vol. 77, pp. 592–599, 2014, doi: 10.1016/j.carbon.2014.05.063.

[51] Worsley, M. A., P. J. Pauzauskie, T. Y. Olson, J. Biener, J. H. Satcher, and T. F. Baumann, "Synthesis of Graphene Aerogel with High Electrical Conductivity," *J. Am. Chem. Soc.*, pp. 14067–14069, 2010, [Online]. Available: http://www.keysight.com/upload/cmc_upload/All/6C0608753-90480-08-12-1999.pdf?&cc=GB&lc=eng.

[52] Worsley, M. A., et al., "High surface area, sp2-cross-linked three-dimensional graphene monoliths," *J. Phys. Chem. Lett.*, vol. 2, no. 8, pp. 921–925, 2011, doi: 10.1021/jz200223x.

[53] Meng, F., X. Zhang, B. Xu, S. Yue, H. Guo, and Y. Luo, "Alkali-treated graphene oxide as a solid base catalyst: Synthesis and electrochemical capacitance of graphene/carbon composite aerogels," *J. Mater. Chem.*, vol. 21, no. 46, pp. 18537–18539, 2011, doi: 10.1039/c1jm13960f.

[54] Fan, X., et al., "Deoxygenation of exfoliated graphite oxide under alkaline conditions: a green route to graphene preparation," *Adv. Mater.*, vol. 20, no. 23, pp. 4490–4493, 2008, doi: 10.1002/adma.200801306.

[55] Vickery, J. L., A. J. Patil, and S. Mann, "Fabrication of graphene-polymer nanocomposites with higher-order three-dimensional architectures," *Adv. Mater.*, vol. 21, no. 21, pp. 2180–2184, 2009, doi:10. 1002/adma.200803606.

[56] Nielsen, T. K., et al., "Nanoconfined NaAlH4: Prolific effects from increased surface area and pore volume," *Nanoscale*, vol. 6, no. 1, pp. 599–607, 2014, doi: 10.1039/c3nr03538g.

[57] Lavoine, N., and L. Bergström, "Nanocellulose-based foams and aerogels: Processing, properties, and applications," *J. Mater. Chem. A*, vol. 5, no. 31, pp. 16105–16117, 2017, doi: 10.1039/c7ta02807e.

[58] Wang, Y., M. D. Gawryla, and D. A. Schiraldi, "Effects of freezing conditions on the morphology and mechanical properties of clay and polymer/clay aerogels," *J. Appl. Polym. Sci.*, vol. 129, no. 3, pp. 1637–1641, 2013, doi: 10.1002/app.39143.

[59] O'Brien, F. J., B. A. Harley, I. V. Yannas, and L. Gibson, "Influence of freezing rate on pore structure in freeze-dried collagen-GAG scaffolds," *Biomaterials*, vol. 25, no. 6, pp. 1077–1086, 2004, doi: 10.1016/S0142-9612(03)00630-6.

[60] Nakazawa, H., H. Yamada, T. Fujita, and Y. Ito, "Texture Control of Clay-Aerogel Through the Crystallization Process of Ice," *Clay Sci.*, vol. 6, no. 6, pp. 269–276, 1987, doi: 10.11362/jcssjclayscience1960.6.269.

[61] Svagan, A. J., P. Jensen, S. V. Dvinskikh, I. Furó, and L. A. Berglund, "Towards tailored hierarchical structures in cellulose nanocomposite biofoams prepared by freezing/freeze-drying," *J. Mater. Chem.*, vol. 20, no. 32, pp. 6646–6654, 2010, doi: 10.1039/c0jm00779j.

[62] Wang, S., Z. Wang, R. Futamura, M. Endo, and K. Kaneko, "Highly microporous-graphene aerogel monolith of unidirectional honeycomb macro-textures," *Chem. Phys. Lett.*, vol. 673, pp. 38–43, 2017, doi: 10.1016/j.cplett. 2017.02.007.

[63] Iijima, S., "Helical microtubules of graphitic carbon," *Nature*, vol. 354, pp. 56–58, 1991.

[64] Saifuddin, A. R., and N. Juniazah, "Carbon Nanotube: A Review on Structure and Their Interaction," *J. Chem.*, vol. 2013, pp. 1–18, 2013.

[65] Meyyappan, M., L. Delzeit, A. Cassell, and D. Hash, "Carbon nanotube growth by PECVD: A review," *Plasma Sources Sci. Technol.*, vol. 12, no. 2, pp. 205–216, 2003, doi: 10.1088/0963-0252/12/2/312.

[66] Wang, R., G. Li, Y. Dong, Y. Chi, and G. Chen, "Carbon quantum dot-functionalized aerogels for NO2 gas sensing," *Anal. Chem.*, vol. 85, no. 17, pp. 8065–8069, 2013, doi: 10.1021/ac401880h.

[67] Bryning, M. B., D. E. Milkie, M. F. Islam, L. A. Hough, J. M. Kikkawa, and A. G. Yodh, "Carbon nanotube aerogels," *Adv. Mater.*, vol. 19, no. 5, pp. 661–664, 2007, doi: 10.1002/adma.200601748.

[68] Li, Y. L., I. A. Kinloch, and A. H. Windle, "Direct Spinning of Carbon Nanotube Fibers from Chemical Vapor Deposition Synthesis," *Science (80-.).*, vol. 304, no. 5668, pp. 276–278, 2004, doi: 10.1126/science.1094982.

[69] Lin, Z., Z. Zeng, X. Gui, Z. Tang, M. Zou, and A. Cao, "Carbon Nanotube Sponges, Aerogels, and Hierarchical Composites: Synthesis, Properties, and Energy Applications," *Adv. Energy Mater.*, vol. 6, no. 17, 2016, doi: 10.1002/ aenm.201600554.

[70] Gui, X., et al., "Carbon nanotube sponges," *Adv. Mater.*, vol. 22, no. 5, pp. 617–621, 2010, doi: 10.1002/adma.200902986.

[71] Kim, K. H., Y. Oh, and M. F. Islam, "Mechanical and thermal management characteristics of ultrahigh surface area single-walled carbon nanotube aerogels," *Adv. Funct. Mater.*, vol. 23, no. 3, pp. 377–383, 2013, doi: 10.1002/adfm. 201201055.

[72] Xia, D., et al., "Tuning the Electrical and Solar Thermal Heating Efficiencies of Nanocarbon Aerogels," *Chem. Mater.*, vol. 33, no. 1, pp. 392–402, 2021, doi: 10.1021/acs.chemmater.0c04166.

[73] Deville, S., "Ice-templating, freeze casting: Beyond materials processing," *J. Mater. Res.*, vol. 28, no. 17, pp. 2202–2219, 2013, doi: 10.1557/jmr.2013.105.

[74] Zou, J., et al., "Ultralight multiwalled carbon nanotube aerogel," *ACS Nano*, vol. 4, no. 12, pp. 7293–7302, 2010, doi: 10.1021/nn102246a.

[75] Greiner, N. R., D. S. Phillips, J. D. Johnson, and F. Volk, "Diamonds in detonation soot," *Nature*, 1988.

[76] Ozawa, M., M. Inaguma, M. Takahashi, F. Kataoka, A. Krüger, and E. Osawa, "Preparation and behavior of brownish, clear nanodiamond colloids," *Adv. Mater.*, vol. 19, no. 9, pp. 1201–1206, 2007, doi: 10.1002/adma.200601452.

[77] Mochalin, V. N., O. Shenderova, D. Ho, and Y. Gogotsi, "The properties and applications of nanodiamonds," *Nat. Nanotechnol.*, vol. 7, no. 1, pp. 11–23, 2012, doi: 10.1038/nnano.2011.209.

[78] Pauzauskie P. J., et al., "Synthesis and characterization of a nanocrystalline diamond aerogel," *Proc. Natl. Acad. Sci. U. S. A.*, vol. 108, no. 21, pp. 8550–8553, 2011, doi: 10.1073/pnas.1010600108.

[79] Manandhar, S., et al., "Rapid sol-gel synthesis of nanodiamond aerogel," *J. Mater. Res.*, vol. 29, no. 24, pp. 2905–2911, 2014, doi: 10.1557/jmr.2014.336.

[80] Biener, J., et al., "Advanced carbon aerogels for energy applications," *Energy Environ. Sci.*, vol. 4, no. 3, pp. 656–667, 2011, doi: 10.1039/c0ee00627k.

[81] Hao, P., et al., "Graphene-based nitrogen self-doped hierarchical porous carbon aerogels derived from chitosan for high performance supercapacitors," *Nano Energy*, vol. 15, pp. 9–23, 2015, doi: 10.1016/j.nanoen.2015.02.035.

[82] Mao, J., J. Iocozzia, J. Huang, K. Meng, Y. Lai, and Z. Lin, "Graphene aerogels for efficient energy storage and conversion," *Energy Environ. Sci.*, vol. 11, no. 4, pp. 772–799, 2018, doi: 10.1039/c7ee03031b.

[83] Luan, V. H., et al., "Synthesis of a highly conductive and large surface area graphene oxide hydrogel and its use in a supercapacitor," *J. Mater. Chem. A*, vol. 1, no. 2, pp. 208–211, 2013, doi: 10.1039/c2ta00444e.

[84] Zhang, L., and G. Shi, "Preparation of highly conductive graphene hydrogels for fabricating supercapacitors with high rate capability," *J. Phys. Chem. C*, vol. 115, no. 34, pp. 17206–17212, 2011, doi: 10.1021/jp204036a.

[85] and R. S. R. Meryl D. Stoller, Sungjin Park, Yanwu Zhu, Jinho An, "Graphene-Based Ultracapacitors," *Nano Lett.*, 2008, doi: 10.1021/acs.langmuir.6b03602.

[86] Li, J., et al., "Structure and electrochemical properties of carbon aerogels synthesized at ambient temperatures as supercapacitors," *J. Non. Cryst. Solids*, vol. 354, no. 1, pp. 19–24, 2008, doi: 10.1016/j.jnoncrysol.2007.07.024.

[87] Schlapbach, L., and A. Züttel, "Hydrogen-storage materials for mobile applications," vol. 414, no. November, pp. 353–358, 2001.

[88] Panella, B., M. Hirscher, and S. Roth, "Hydrogen adsorption in different carbon nanostructures," *Carbon N. Y.*, vol. 43, no. 10, pp. 2209–2214, 2005, doi: 10.1016/j.carbon.2005.03.037.

[89] Chahine, R., and P. Bénard, "*Adsorption Storage of Gaseous Hydrogen at Cryogenic Tempera Tures*," vol. 43, pp. 1257–1264, 1998.

[90] Wang, J., and S. Kaskel, "KOH activation of carbon-based materials for energy storage," *J. Mater. Chem.*, vol. 22, no. 45, pp. 23710–23725, 2012, doi: 10.1039/c2jm34066f.

[91] "*Sequía en México: 4 imágenes satelitales que muestran los graves efectos de la escasez de lluvias - BBC News Mundo*." https://www.bbc.com/mundo/noticias-america-latina-57163233 (accessed Jul. 10, 2021).

[92] Welgemoed, T. J., and C. F. Schutte, "Capacitive Deionization Technology™: An alternative desalination solution," *Desalination*, vol. 183, no. 1–3, pp. 327–340, 2005, doi: 10.1016/j.desal.2005.02.054.

[93] Thamilselvan, A., A. S. Nesaraj, and M. Noel, "Review on carbon-based electrode materials for application in capacitive deionization process," *Int. J. Environ. Sci. Technol.*, vol. 13, no. 12, pp. 2961–2976, 2016, doi: 10.1007/ s13762-016-1061-9.

[94] Jung, H. H., S. W. Hwang, S. H. Hyun, K. H. Lee, and G. T. Kim, "Capacitive deionization characteristics of nanostructured carbon aerogel electrodes synthesized via ambient drying," *Desalination*, vol. 216, no. 1–3, pp. 377–385, 2007, doi: 10.1016/j.desal.2006.11.023.

[95] Storer, D. P., J. L. Phelps, X. Wu, G. Owens, N. I. Khan, and H. Xu, "Graphene and Rice-Straw-Fiber-Based 3D Photothermal Aerogels for Highly Efficient Solar Evaporation," *ACS Appl. Mater. Interfaces*, vol. 12, no. 13, pp. 15279–15287, 2020, doi: 10.1021/acsami.0c01707.

[96] Li, R., C. Chen, J. Li, L. Xu, G. Xiao, and D. Yan, "A facile approach to superhydrophobic and superoleophilic graphene/polymer aerogels," *J. Mater. Chem. A*, vol. 2, no. 9, pp. 3057–3064, 2014, doi: 10.1039/c3ta14262k.

[97] Wu, T., et al., "Three-dimensional graphene-based aerogels prepared by a self-assembly process and its excellent catalytic and absorbing performance," *J. Mater. Chem. A*, no. 207890, p. 121, 2013.

[98] Li, J., et al., "Ultra-light, compressible and fire-resistant graphene aerogel as a highly efficient and recyclable absorbent for organic liquids," *J. Mater. Chem. A*, vol. 2, no. 9, pp. 2934–2941, 2014, doi: 10.1039/c3ta14725h.

[99] Gui, X., et al., "Magnetic and highly recyclable macroporous carbon nanotubes for spilled oil sorption and separation," *ACS Appl. Mater. Interfaces*, vol. 5, no. 12, pp. 5845–5850, 2013, doi: 10.1021/am4015007.

[100] Xiao, B., and K. M. Thomas, "Competitive adsorption of aqueous metal ions on an oxidized nanoporous activated carbon," *Langmuir*, vol. 20, no. 11, pp. 4566–4578, 2004, doi: 10.1021/la049712j.

[101] Zhuyin, S., M. Qinghan, Z. Xuetong, M. Rui, and C. Bing, "Green synthesis of carbon nanotube–graphene hybrid aerogels and their use as versatile agents for water purification," no. Vc, pp. 8767–8771, 2012, doi: 10.1039/c2jm0005e.

[102] Yin, H., C. Zhang, F. Liu, and Y. Hou, "Hybrid of iron nitride and nitrogen-doped graphene aerogel as synergistic catalyst for oxygen reduction reaction," *Adv. Funct. Mater.*, vol. 24, no. 20, pp. 2930–2937, 2014, doi: 10.1002/adfm. 201303902.

[103] Zhang, X., D. Liu, L. Yang, L. Zhou, and T. You, "Self-assembled three-dimensional graphene-based materials for dye adsorption and catalysis," *J. Mater. Chem. A*, vol. 3, no. 18, pp. 10031–10037, 2015, doi: 10.1039/c5ta00355e.

[104] Dubey, S. P., A. D. Dwivedi, I. C. Kim, M. Sillanpaa, Y. N. Kwon, and C. Lee, "Synthesis of graphene-carbon sphere hybrid aerogel with silver nanoparticles and its catalytic and adsorption applications," *Chem. Eng. J.*, vol. 244, pp. 160–167, 2014, doi: 10.1016/j.cej.2014.01.042.

[105] Shen, Y., et al., "A high-capacity lithium-air battery with Pd modified carbon nanotube sponge cathode working in regular air," *Carbon N. Y.*, vol. 62, pp. 288–295, 2013, doi: 10.1016/j.carbon.2013.05.066.

[106] Qi, H., J. Liu, J. Pionteck, P. Pötschke, and E. Mäder, "Carbon nanotube-cellulose composite aerogels for vapour sensing," *Sensors Actuators, B Chem.*, vol. 213, pp. 20–26, 2015, doi: 10.1016/j.snb.2015.02.067.

[107] Long H., et al., "High Surface Area MoS2/Graphene Hybrid Aerogel for Ultrasensitive NO2 Detection," *Adv. Funct. Mater.*, vol. 26, no. 28, pp. 5158–5165, 2016, doi: 10.1002/adfm.201601562.

[108] Juanjuan, Z., L. Ruiyi, L. Zaijun, L. Junkang, G. Zhiguo, and W. Guangli, "Synthesis of nitrogen-doped activated graphene aerogel/gold nanoparticles and its application for electrochemical detection of hydroquinone and o-dihydroxybenzene," *Nanoscale*, vol. 6, no. 10, pp. 5458–5466, 2014, doi: 10.1039/c4nr00005f.

[109] Xu, X. et al., "Self-sensing, ultralight, and conductive 3D graphene/iron oxide aerogel elastomer Deformable in a Magnetic Field," *ACS Nano*, vol. 9, no. 4, pp. 3969–3977, 2015, doi: 10.1021/nn507426u.

[110] Wan, Y. J.. P. L. Zhu, S. H. Yu, R. Sun, C. P. Wong, and W. H. Liao, "Ultralight, super-elastic and volume-preserving cellulose fiber/graphene aerogel for high-performance electromagnetic interference shielding," *Carbon N. Y.*, vol. 115, pp. 629–639, 2017, doi: 10.1016/j.carbon.2017.01.054.

[111] Foroughi, J. et al., "Knitted Carbon-Nanotube-Sheath/Spandex-Core Elastomeric Yarns for Artificial Muscles and Strain Sensing," *ACS Nano*, vol. 10, no. 10, pp. 9129–9135, 2016, doi: 10.1021/acsnano.6b04125.

Chapter 8

An International Week Event as an Approach to Introduce Engineering Students to Graphene Nanotechnology

Alina I. Pruna[1,2,*]
[1]Institute of Materials Technology, Universitat Politecnica de Valencia, Valencia, Spain
[2]Center for Surface Science and Nanotechnology, University Politehnica of Bucharest, Bucharest, Romania

Abstract

One of the top-priority focuses in education is Nanotechnology (nanoscale science and engineering, as well as nano-enabled technologies) due to its great impact on society through the benefits in energy saving, effective use of raw materials, environmental protection or medical advances, which are all connected to engineering. Graphene nanomaterials have emerged as a highly attractive alternative and their aerogel assemblies are investigated in a large number of applications including medical, water treatment or CO_2 capture. As nanomaterials in general and graphene nanomaterials in particular are prone to become viable for manufacturing processes, the need for skilled and extensively informed workforce increases. To this aim, the engineering students need to be introduced to nanotechnology and nanotechnology focused on graphene as study case, in order to help developing their abilities and improve curriculum. In this short chapter, the case study of an International Week event based on the participation of guest speakers from industry and research centers is considered as an approach to introduce the second- and third-year engineering students to Nanotechnology in order to improve their education and training. Student

---

* Corresponding Author's Email: apruna@itm.upv.es.

In: Aerogels: Properties and Applications
Editor: Alina Iuliana Pruna
ISBN: 978-1-68507-788-4

learning was assessed by post-testing and satisfaction surveys. Results indicated that creating visual elements, videos and solving real life issues improved their learning. Moreover, the obtained results indicated an International Week event may provide an excellent approach to integrate Nanotechnology and Graphene Nanotechnology in the curriculum framework for exposing the students to Nanotechnology by sharing the research and industry professionals´ experience.

Keywords: engineering, graphene, nanotechnology, teaching, international week

Introduction

The number of manufactured goods based on incorporated nanotechnology has been observed to increase exponentially and in the longer term, it is expected that nanotechnologies will provide more efficient approaches to manufacturing by using less raw materials and energy. Moreover, graphene nanomaterial research has been expanding greatly over the last decade, since it offers excellent properties including thermal, mechanical, electrical, optical etc. It was observed that lighter and stronger materials are greatly researched and desired to be used in different fields including aircraft, electronics, energy etc., since they provide a reduction in size of equipment, decrease in energy consumption, and an increase in energy production efficiency. In the future, the energy consumption will get decreased and more smaller devices are expected which are more convenient to use. The 3-dimensional porous assemblies of graphene attracted increased interest in a variety of fields including wastewater treatment as a sorbent material or air cleaning as a CO_2 capture material.

It is acknowledged that nanotechnology is a multidisciplinary and highly competitive field that makes one of the most highly funded fields. Given the important role of nanotechnology in the future development, important efforts are being dedicated to provide highly qualified and well-trained manpower to satisfy the industry needs. Nanotechnology is a very fast developing field that requires a lot of creative ideas besides the understanding of its concepts. The concept of nanotechnology should be introduced in the curriculum in order to keep the pace with this powerful technology. Each engineering program should revise their curriculum content in order to ensure the exposure of students to elements of nanotechnology and also train the lecturers in this respect. As early the exposure to nanotechnology, the greater the benefit for the students.

Integrated approaches to Nanotechnology education are sought to help the next generation students to solve real-world problems to helping them to develop abilities including critical thinking or creativity [1]. There have been several attempts to the nanotechnology integration to engineering curriculum by the researchers, for example, Greenberg [2] stated the undergraduated level has less barriers to integration, Ghattas and Carver [3] argued on the integration from the school level. Later, Malsch [4] identified European employers' needs for nano-education. Nevertheless, a large gap between research and application in education was observed, where a main reason is constituted by the lack of understanding of nanotechnology.

The aim of this contribution is to give a short insight into the student perception on the use of International Week event as an approach to introduce the students to Nanotechnology. Given the challenge for educators to integrate Nanotechnology due to their limited training in this field, the International Week event was taken into consideration for this purpose since there are teachers and researchers with varying formation applying for mobility training and teaching at this event. In this experience, a short Nanotechnology course was offered by an experienced researcher with expertise in nanotechnology related fields using the International Week event as an approach to maximize students' exposure to a greater variety of examples, despite a potential barrier of teaching abilities. As subjects for the Nanotechnology course, porous materials have been highlighted, specifically ZnO and graphene aerogels. To the best of author's knowledge, no data has been reported so far investigating the engineering students' perceptions of nanotechnology in these conditions.

The short course on Nanotechnology introduced some scientific principles, synthesis and characterization equipment and related applications. It provided students with a basic understanding of how nanotechnology and nanomaterials really work. The focus was centred on the synthesis, characterization, and application of porous ZnO films and graphene aerogels.

The chapter makes a short overview on Nanotechnology courses that are offered to engineering students, the methodology employed to structure and delivery of the course to engineering students together with the analysis of results, the discussion on student satisfaction results and conclusions.

Courses on Nanotechnology

The concepts of nano were first introduced by an American physicist Richard Feynman [5] with his speech 'There's Plenty of Room at the Bottom' in 1959, followed by Taniguchi in 1974 mentioning the term nanotechnology. This term is used to refer to materials in the scale from 1 to 100 nm [6, 7]. The materials exhibit different properties at the nanoscale than at the macroscale and exploiting the nanoscaled materials to obtain superior properties has become greatly employed approach in different fields of application. Given the emergence of nanotechnology as a research field lately, it was concluded that engineering students need to gain nanotechnology skills required by the industry [8].

In the last years, many universities worldwide have started to provide the engineering undergraduates with exposure to courses on Nanotechnology. Table 1 shows the aims of typical courses on Nanotechnology Fundamentals. This introductory course could require as a pre-requisite only pre-university biology, chemistry or physics and thus can be offered in year 1 or 2 within a 4-year curriculum.

Table 1. Outline of typical course on Nanotechnology Fundamentals

Title/Scope	Description
Introduction to Nanotechnology	What is nanotechnology The scale of materials Nanotechnology applications and products Impact of nanotechnology on society
Nanostructure of the universe	Atomic and electronic structure Types of chemical bonds and bonding Periodic table and chemical properties
Nanomaterial properties	Surface area Quantum computing Nanoparticles vs micron-sized particles
Characterization methods	Material characterization techniques and equipment
Synthesis approaches	Top-Down and Bottom-Up processing methods
Examples of nanotechnology	Devices
Social / ethical issues	Dispersion in environment

On the other hand, Table 2 shows a short selection of such courses offered by many universities worldwide. As it can be observed there are many nanotechnology-related courses offered to engineering students to help them increase their related skills.

Table 2. Selection of Nanotechnology courses offered by several universities to different fields of engineering

Specialization	Courses offered
Chemical Engineering	Micro/Nano Processing Technology (MIT, US) Nanostructure of materials (Monash, Australia) Nanotechnology in chemical engineering (Xiamen University, China)
Mechanical Engineering	Introduction to Nano/Microfabrication (Princeton, US)
Electrical and Electronic Engineering	Nanoscale Science & Engineering (Penn. University, US) Microelectronics and Nanofabrication (Leeds University, UK)
Civil Engineering	Environmental Applications of Nanotechnology (University of Western Ontario, Canada) Environmental Nanoscience (Duke U)

Methodology

Course Exposure for Engineering Students

In table 3 the engineering fields to which the Nanotechnology course proposal was introduced to are presented. These specialization fields were the same over the 4-year experience. It can be observed that the course was offered to both 2 and 3-year course students.

Table 3. Engineering specializations and student levels exposed to the course

Specialization	Year
Mechanical Engineering	2 / 3
Industrial Electronics and Automation Engineering	2 / 3

Teaching Model

The short Nanotechnology course was an 'optional' one in the engineering students' curriculum. The students selected this course voluntarily from those offered during the International Week event at Florida Universitaria, Valencia, Spain in the period 2018-2021.

The teaching methodology applied in the course is further shown on in scheme 1. As it can be observed in scheme 1, it followed a hybrid model based on traditional lectures, images, videos, sampling of products and scientific article/patent discussions to make the acquisition of new knowledge more interesting and motivating. A traditional laboratory-theory course usually presents concepts during a lecture session in classroom and then students learn applications of those concepts in a laboratory practice. However, in this course, the lectures were only briefly presenting the basic concepts and focusing more on samples of nanotechnology-based products and their application.

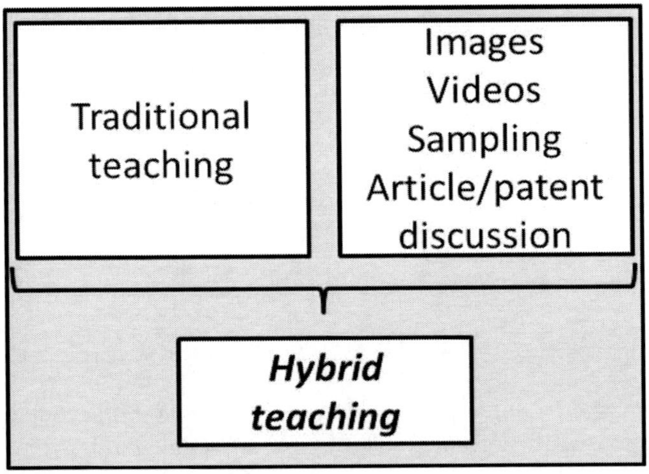

Figure 1. Teaching model for Nanotechnology course during International Week event.

The hypothesis for this work was that student learning in nanotechnology would be improved by visual illustrations such as images, videos and sampling of products and thus be more motivated [9-11]. Several hands-on activities were used in the classroom, including measuring dimensions (several examples on graphene and ZnO nanomaterials in micrographs such as scanning electron and transmission electron micrographs) or estimating apparent density of nanomaterial-based products (samples of graphene aerogels). This promoted interest and motivation, as the students realized how nanomaterials can be obtained from different sources and the products they can achieve. The impact of the course on learning outcomes was evaluated by a post-course satisfaction survey when the students responded on 5-point Likert scale.

Learning Objectives

The short course on Nanotechnology offered during the International Event was based on the educational outcomes introduced by Uddin to be expected from a nanotechnology course [12]:

- 'Provide understanding, characterization, and measurements of nanostructure properties'
- 'Provide ability for synthesis, processing and manufacturing of nanocomponents and nanosystems'
- 'Provide ability for design, analysis and simulation of nanostructures and nanodevices'
- 'Prepare students to conduct research and development of economically feasible and innovative applications of nanodevices in all spheres of our daily life'

Course Content

The course had a duration of 10 h and it corresponded with 1 ECTS upon finalizing the course. The short course on Nanotechnology with graphene aerogels as study case was proposed with the following content:

1. General introduction to nanoscience and nanotechnology
2. Nanoscale. Effects of size and form. Nanoscale and macroscale level examples.
3. The importance of size effect in nanotechnology. Study case: ZnO nanowires
4. Carbon nanomaterials. Study case: Graphene

5. Synthesis of graphene nanomaterials: scotch-tape and wet chemistry study cases
6. Synthesis of graphene 3D assemblies – Study case: hydrothermal approach
7. Characterization methods of nanoscale materials – scanning electron miscropy, transmission electron microscopy, X-ray diffraction, Raman spectroscopy.
8. Applications of nanotechnology – case of ZnO and graphene porous assemblies, graphene-based aerogels in photocatalytic treatment of wastewaters, cosmetics, energy storage, oil absorption and CO_2 capture.
9. Identifying risks in nanotechnology
10. Discussion on scientific articles and patent in graphene aerogels

Briefly, understanding the nanoscale is one concept that students need to grasp [13]. Students are introduced to the concept of Nanotechnology as a multidisciplinary field concerned with the control of matter in the 1-100 nanoscale. At nanoscale, the materials may have different molecular organizations and properties compared to the same material on macroscale. These unique properties indicate that nanoscale materials can behave differently in the human body and in the environment than the corresponding macro-scale counterparts [12, 14, 15]. These unique properties enable nanotechnologies to be used to optimize a variety of product.

Furthermore, it is important to connect the nanomaterials to their applications since it was observed there is a lack in this respect [16] and to learn about nanoscale measurement instruments.

Emphasis has been put on ZnO porous films and graphene aerogels. ZnO is well known to exhibit a high versatility of morphologies and many techniques can be applied to synthesize porous ZnO films. As an example, the electrodeposition methods were shown as these can offer the user the control over the coverage degree, aspect ratio of nanostructures and it can be applied to coat a large variety of support shapes that are difficult by other methods. On the other hand, graphene nanomaterials are an excellent topic to show to engineering students. Starting with scotch tape methods, that granted a Nobel prize and up to wet-chemistry approach, the graphene nanomaterials can be achieved by simple methods. Images and videos related to the synthesis of both ZnO and graphene nanomaterials have been shown. Moreover, as a nanomaterial source to achieve macro-assemblies, the graphene oxide (the oxidized form of graphite) has been exemplified to obtain graphene aerogels by a simple, cheap method, namely hydrothermal one. By simply using a reactor and an oven, students have been shown how to obtain monoliths of graphene, that is graphene oxide aerogels, by firstly achieving the hydrogel form and further in the form of aerogel, by removal of the solvent in the hydrogel. Such assemblies exhibit very low density and excellent mechanical properties. Samples of graphene aerogels have been offered to the students for visual inspection, and apparent density measurement by simply weighing the aerogel with a pocket scale and measuring their geometric dimensions with a caliper.

Student Learning Expectations

The short course on Nanotechnology offered during International Week event to the engineering students intended the students to improve their level of understanding of nanotechnology knowledge upon completion of the course. After the course, the student would be able to:

- Develop vocabulary to be used in nanotechnology
- Have a conceptual understanding of the nanometer scale
- Explain how nanotechnology works as it pertains to real-world applications
- Describe graphene can be obtained by scoth tape and wet chemistry methods
- Describe how ZnO porous films can be obtained
- Name and briefly describe the characterization equipments employed for nanomaterials
- Identify some applications and potential applications for specific graphene nanotechnologies
- Describe water treatment by photocatalytic degradation and oil spill absorption with graphene-zno based aerogels
- Discuss the safety issues related to graphene nanomaterials and patenting the scientific results

Transversal Competencies

The expectations of the course are envisioned to be achieved as the students are expected to develop transversal competencies, as shown in table 4. The means offered during the course for reaching such abilities is detailed in each case of transversal competency. As it can be observed, the English language and technical vocabulary were meant to be developed by using English as delivery language bot only for presenting the course but also for studying scientific articles/patents.

Different fields of application were exemplified in both the course and the selected articles/patents, and the students were asked to comment on the efficiency of the nanomaterials by using mainly the abstract section of these materials.

The course also included a section dedicated to developing the innovation, creativity, and entrepreneurship, where the students have been introduced to the differences between the academic research and industry and compared the scientific articles with patents.

Since the students were divided in groups to study the articles and patents, the teamwork abilities were also envisioned. Upon thc study of the articles and patents, the students were asked to briefly present the work, by highlighting the type of nanomaterial employed, the characterization equipment, the field of application and material performance, thus contributing to developing the communication skills.

Table 4. Transversal competencies to develop upon finalizing the course

Transversal competency	**Means**
Compression and integration	Improvement of English Understanding scientific articles
Application and practical thinking	Brainstorming on energy efficiency problems. Search for solutions
Innovation, creativity, and entrepreneurship	Academic research vs industry
Teamwork and leadership	Team analysis of a scientific article
Ethical, environmental responsibility	Analysis of sustainable solutions to real energy problems
Effective communication	Oral presentation of the results of teamwork
Critical thinking	Analysis of scientific articles based on acquired knowledge

Student Assessment

At the end of the short course on Nanotechnology offered during International Week event, the engineering students were exposed to scientific articles and patents related to nanotechnology. The students were divided in groups of 2 to 3 students and asked to identify the characterization methods and equipments employed in the articles, as introduced in the course, the application field of graphene nanomaterials and the subjects of patents. The students would use vocabulary and terms specific to nanotechnology. The activity was obligatory to pass the course and pass/fail grading was employed.

Student Satisfaction Survey

The engineering students' perception on the course on nanotechnology was investigated through their voluntary responses to a post-course satisfaction survey. By using a Likert scale (1 – 5 scoring system), the students were asked to answer questions that were scored as: 5 = strongly agree, 4 = agree, 3 = maybe, 2 = disagree, and 1 = strongly disagree.

Results and Discussion

The quantitative results are presented in this section in tables and figures. Data analysis was used to answer varying questions including the perception of nanotechnology, in terms of exposure, motivation etc. Given that the data were ordinal, the mode rather than the mean score was used – see table 5-7 for the results obtained in 2018, as exemplification. Varying aspects of nanotechnology were covered, namely nanoexposure, nanoawareness and nanomotivation.

Table 5. Mode scores received on aspects related to nanoexposure

Nanoexposure	Mode score (1 - strongly disagree, 5- strongly agree)
There has been theoretical nanotechnology input	4
Read something about nanotechnology	5
Use of videos and images on nanotechnology	4
Samples offered	5

Table 6. Mode scores received on aspects related to nanoawareness

Nanoawareness	Mode score (1 -strongly disagree, 5- strongly agree)
There have been field of applications	5
Nanoscale objected presented	5
Methods of synthesis presented	5

Table 7. Mode scores received on aspects related to nanomotivation

Nanomotivation	Mode score (1 -strongly disagree, 5- strongly agree)
Watch content of nanotechnology	5
Give a presentation in nanotechnology	5
I find the contents useful	5
The teacher created good communication	5
The class atmosphere was	5
The course duration was	4
My motivation was	4
My participation was	5
The contribution to expand my knowledge of the topic was	5

All the aspects – nanoexposure, nanoawareness and nanomotivation were covered in the survey by varying statements, to which the students were invited to agree on the 5-point Likert scale. As it can be observed, tables 5-7 provide the detailed descriptive statistics regarding the participants' exposure, awareness, and motivation in regard to the Nanotechnology course offered. All aspects received mode scores above 4. It is evident that the students were presented brief theoretical aspects of nanotechnology and they also agreed that images, videos, products, and application cases were beneficial in learning about nanotechnology. This can be attributed to the fact that the students have insufficient knowledge about nanotechnology, their information being limited to internet, TV or other products and as such, a course on Nanotechnology, even as a short seminar of 10 h, it is useful to introduce them to the field. Thus, the teaching model was evaluated as very good.

Further, Fig. 2 presents the evolution of student satisfaction regarding the course of Nanotechnology over the period of implementation of the project, namely the Introduction to Nanotechnology during International Week events in th period 2018-2021. The students found the course motivating under different aspects as indicated by the scores received for each item being close to or above 4. More precisely, an increase in the score was observed after the first year, which is attributed to an improved adaptation of the course materials to the students needs, in agreement with the first survey. The slight decrease afterwards might be correlated to the advances in nanotechnology that are more present in the media lately, and thus, the students get more insights into the topic. Another issue could be the limited capacity of the instructor to deliver teaching content, as this belonged to academia research and not to the academia teaching.

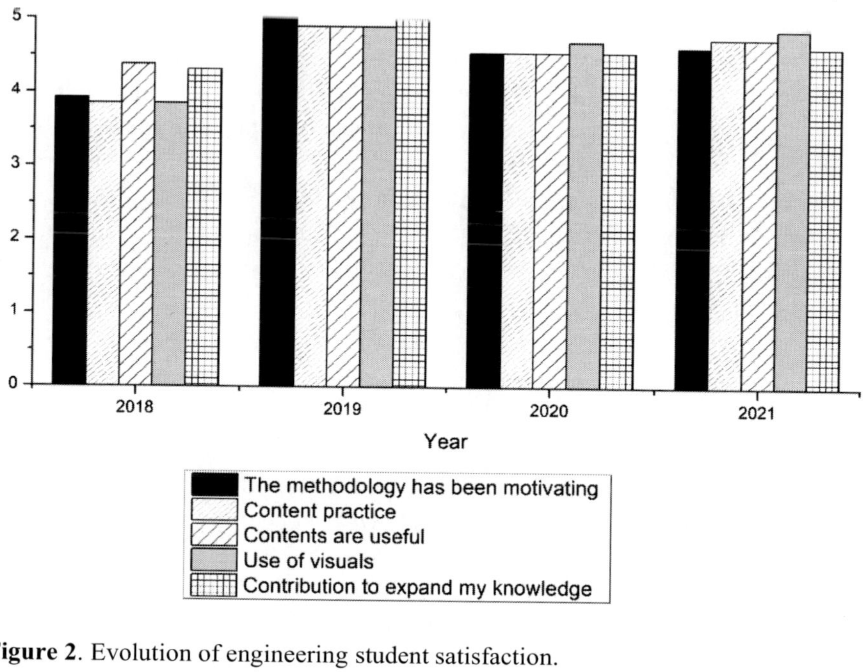

Figure 2. Evolution of engineering student satisfaction.

Conclusion

Given the growing interest in, and relevance of integrated approaches to Nanotechnology education, there is an urgent desire to understand the challenges and obstacles in developing and implementing integrated Nanotechnology curriculum and instruction.

This chapter presents an experience of exposing engineering students to a short introductory course in Nanotechnology in the period 2018-2021 by using an International Week event as means of integration. The effectiveness of such approach in fostering student motivation and engagement as well as disciplinary content learning was indicated by the scores received over the years of implementation and the positive evolution. However, is it considered that the teaching hypothesis for student learning would be further improved by using different participative activities related to nanotechnology.

Future Perspectives

In order to efficiently integrate Nanotechnology into student education there is a need to strengthen teachers' content area knowledge and classroom management skills as they are needed to efficiently run integrated lessons. Furthermore, instructors need to develop pedagogical skills that encourage students to collaborate which may require interdisciplinary teams of instructors and to provide with the required opportunities to interdisciplinary teams to work together. Moreover, teachers also need to promote teamwork and communication skills.

References

[1] Burrows A., T. Slater, A proposed integrated STEM framework for contemporary teacher preparation. *Teacher Education and Practice*, 28 (2015) 318 – 330.

[2] Greenberg A., Integrating nanoscience into the classroom: Perspectives on nanoscience education projects. *Acs Nano* 3 (2009) 762-769.

[3] Ghattas N. I., J. S. Carver, Integrating nanotechnology into school education: A review of the literature. *Research in Science & Technological Education* 30 (2012) 271-284.

[4] Malsch I., Nano-education from a european perspective : Nano-training for non-r&d jobs. *Nanotechnology Reviews* 3 (2014) 211-221.

[5] Taniguchi N., "On the Basic Concept of 'Nano-Technology'," Proc. Intl. Conf. Prod. Eng. Tokyo, Part II, *Japan Society of Precision Engineering*, 1974.

[6] Blonder R., S. Sakhnini, Teaching two basic nanotechnology concepts in secondary school by using a variety of teaching methods. *Chemistry Education Research and Practice* 13 (2012) 500-516.
[7] Ramsden J. J., Chapter 1 - what is nanotechnology? In J. J. Ramsden (Ed.), *Applied nanotechnology* (second edition) (2014) 3-12, Oxford: William Andrew Publishing.
[8] Roco M. C., Converging science and technology at the nanoscale: Opportunities for education and training. *Nature Biotechnology*, 21 (2003) 1247-1249.
[9] Ball D. L., L. Sleep, T. A. Boerst, H. Bass, Combining the development of practice and the practice of development in teacher education. *The Elementary School Journal*, 109 (2009) 458 – 474.
[10] Borko H., J. Jacobs, E. Eiteljorg, M. Pittman, Video as a tool for fostering productive discussions in mathematics professional development. *Teaching and Teacher Education*, 24 (2008) 417 – 436.
[11] Roth K. J., H. E. Garnier, C. Chen, M. Lemmens, K. Schwille, N. I. Z. Wickler, Videobased lesson analysis: Effective science PD for teacher and student learning. *Journal of Research in Science Teaching*, 48 (2011) 117 – 148.
[12] Uddin M., A. R. Chowdhury, Integration of Nanotechnology into the Undergraduate Engineering Curriculum. *International Conference on Engineering Education, Session* 8B2-9 (2001) 6-9.
[13] Fazzaro D. E., D. Newberry, W. Trybula, J. Hyder, Introducing a nanotechnology curriculum and considerations for bridging academic/industry relationships: An overview and the new challenge for ATMAE, *The Journal of Technology, Management, and Applied Engineering* 28 (2012) 1-9.
[14] Sellers K., C. Mackay, L. L. Bergeson, S. R. Clough, M. Hoyt, J. Chen, K. Henry, J. Hamblen, *Nanotechnology and the Environment*. Taylor & Francis Group: CRC Press (2009) Chapters 1 & 2.
[15] Savolainen K., et al. Nanotechnologies, engineered nanomaterials and occupational health and safety – A review. *Safety Science* 48 (2010) 957-963.
[16] Lu K., A study of engineering freshman regarding nanotechnology understanding, *Journal of STEM Education* 10 (2009) 7-16.

Editor's Contact Information

Alina I. Pruna
Institute of Materials Technology
Universitat Politecnica de Valencia
Valencia, Spain
Center for Surface Science and Nanotechnology
University Politechnica of Bucharest
Bucharest, Romania
apruna@itm.upv.es

List of Contributors

Cristina E. Almeida-Naranjo
Department of Mechanical Engineering
Escuela Politécnica Nacional
Quito, Ecuador

Victor H. Guerrero
Department of Materials
Escuela Politécnica Nacional
Quito, Ecuador

Nelly Ma. Rosas-Laverde
Department of Materials
Escuela Politécnica Nacional
Quito, Ecuador

Alina I. Pruna
Institute of Materials Technology
Universitat Politecnica de Valencia
Valencia, Spain
Center for Surface Science and Nanotechnology
University Politechnica of Bucharest
Bucharest, Romania

Shikha Bansal
Centre of Nanoscience and Nanotechnology
Jamia Millia Islamia, New Delhi, India
Waseem Ashraf
Centre of Nanoscience and Nanotechnology
Jamia Millia Islamia, New Delhi, India

Honey Mittal
Centre of Nanoscience and Nanotechnology
Jamia Millia Islamia, New Delhi, India

Manika Khanuja
Centre of Nanoscience and Nanotechnology
Jamia Millia Islamia, New Delhi, India

Norma Gisela Corral Capulin
Dirección de Ingeniería en Biotecnología
Universidad Politécnica del Valle de Toluca
Almoloya de Juárez, Estado de México

Misael García Ruiz
Universidad Autónoma del Estado de México
Centro Conjunto de Investigación en Química Sustentable
UAEM-UNAM

Salomé Gabriela Galeas Hurtado
Materials Department
Escuela Politécnica Nacional
Quito, Ecuador

Jair A. Yánez Sánchez
Department of Materials
Escuela Politécnica Nacional
Quito, Ecuador

Index

A

B

C

D

E

V

W

X